Anatomy & Physiology Workbook

FOR DUMMIES®
A Wiley Brand

2nd Edition

by Janet Rae-Dupree and
Pat DuPree

FOR DUMMIES®
A Wiley Brand

Anatomy & Physiology Workbook For Dummies® 2nd Edition

Published by: **John Wiley & Sons, Inc.,** 111 River Street, Hoboken, NJ 07030-5774, www.wiley.com

Copyright © 2015 by John Wiley & Sons, Inc., Hoboken, New Jersey

Published simultaneously in Canada

For general information on our other products and services, please contact our Customer Care Department within the U.S. at 877-762-2974, outside the U.S. at 317-572-3993, or fax 317-572-4002. For technical support, please visit www.wiley.com/techsupport.

Wiley publishes in a variety of print and electronic formats and by print-on-demand. Some material included with standard print versions of this book may not be included in e-books or in print-on-demand. If this book refers to media such as a CD or DVD that is not included in the version you purchased, you may download this material at http://booksupport.wiley.com. For more information about Wiley products, visit www.wiley.com.

Library of Congress Control Number: 2014946673

ISBN 978-1-118-94007-5 (pbk); ISBN 978-1-118-94008-2 (ebk); ISBN 978-1-118-94009-9 (ebk)

Manufactured in the United States of America

10 9 8 7 6 5 4 3 2 1

Contents at a Glance

Table of Contents

Introduction

Whether your aim is to become a physical therapist or a pharmacist, a doctor or an acupuncturist, a nutritionist or a personal trainer, a registered nurse or a paramedic, a parent or simply a healthy human being — your efforts have to be based on a good understanding of anatomy and physiology. But knowing that the knee bone connects to the thigh bone (or does it?) is just the tip of the iceberg. In *Anatomy & Physiology Workbook For Dummies,* 2nd Edition, you discover intricacies that will leave you agog with wonder. The human body is a miraculous biological machine capable of growing, interacting with the world, and even reproducing despite any number of environmental odds stacked against it. Understanding how the body's interlaced systems accomplish these feats requires a close look at everything from chemistry to structural mechanics.

Early anatomists relied on dissections to study the human body, which is why the Greek word *anatomia* means "to cut up or dissect." Anatomical references have been found in Egypt dating back to 1600 B.C., but it was the Greeks — Hippocrates, in particular — who first dissected bodies for medical study around 420 B.C. That's why more than two millennia later we still use words based on Greek and Latin roots to identify anatomical structures.

That's also part of the reason so much of the study of anatomy and physiology feels like learning a foreign language. Truth be told, you are working with a foreign language, but it's the language of you and the one body you're ever going to have.

About This Book

This workbook isn't meant to replace a textbook, and it's certainly not meant to replace going to an actual anatomy and physiology class. It works best as a supplement to your ongoing education and as a study aid in prepping for exams. That's why we give you insight into what your instructor most likely will emphasize as you move from one body system or structure to the next.

Your coursework most likely will cover things in a different order than we've chosen for this book. We encourage you to take full advantage of the table of contents and the index to find the material addressed in your class. Whatever you do, certainly don't feel obligated to go through this workbook in any particular order. However, please do answer the practice questions and check the answers at the end of each chapter because, in addition to answers, we clarify why the right answer is the right answer and why the other answers are incorrect; we also provide you with memory tools and other tips whenever possible.

Within this book, you may note that some web addresses break across two lines of text. If you're reading this book in print and want to visit one of these web pages, simply key in the web address exactly as it's noted in the text, pretending as though the line break doesn't exist. If you're reading this as an e-book, you've got it easy — just click the web address to be taken directly to the web page.

Foolish Assumptions

In writing *Anatomy & Physiology Workbook For Dummies,* 2nd Edition, we had to make some assumptions about you, the reader. If any of the following apply, this book's for you:

- ✔ You're an advanced high school student or college student trying to puzzle out anatomy and physiology for the first time.
- ✔ You're a student at any level who's returning to the topic after some time away, and you need some refreshing.
- ✔ You're facing an anatomy and physiology exam and want a good study tool to ensure that you have a firm grasp of the topic.

Because this is a workbook, we had to limit our exposition of each and every topic so that we could include lots of practice questions to keep you guessing. (Believe us, we could go on forever about this anatomy and physiology stuff!) In leaving out some of the explanation of the topics covered in this book, we assume that you're not just looking to dabble in anatomy and physiology and therefore have access to at least one textbook on the subject.

Icons Used in This Book

Throughout this book, you'll find symbols in the margins that highlight critical ideas and information. Here's what they mean:

The Tip icon gives you juicy tidbits about how best to remember tricky terms or concepts in anatomy and physiology. It also highlights helpful strategies for fast translation and understanding.

The Remember icon highlights key material that you should pay extra attention to in order to keep everything straight.

The sizzling bomb icon — otherwise known as the Warning icon — points out areas and topics where common pitfalls can lead you astray.

The Example icon marks questions for you to try your hand at. We give you the answer straightaway to get your juices flowing and your brain warmed up for more practice questions.

Beyond the Book

In addition to the material in the print or e-book you're reading right now, this product also comes with some access-anywhere goodies on the web. While it's important to study each anatomical system in detail, it's also helpful to know how to decipher unfamiliar anatomical terms the first time you see them. Check out the free Cheat Sheet at www.dummies. com/cheatsheet/anatomyphysiologywb for a list of the more common Latin and Greek roots, prefixes, and suffixes that will have you telling your gastronomic from your autonomic in no time. We also summarize the cell cycle for you and remind you how to direct your proximal attention away from your distal (in other words, we explain anatomic positions and planes).

If you'd like to dig into information on new anatomy discoveries, the human microbiome, assisted reproduction, and more, check out free articles at www.dummies.com/extras/ anatomyphysiologywb.

Where to Go from Here

If you purchased this book and you're already partway through an anatomy and physiology class, check the table of contents and zoom ahead to whichever segment your instructor is covering currently. When you have a few spare minutes, review the chapters that address topics your class already has covered. It's an excellent way to prep for a midterm or final exam.

If you haven't yet started an anatomy and physiology class, you have the freedom to start wherever you like (although we suggest that you begin with Chapter 1) and proceed onward and upward through the glorious machine that is the human body!

Part I
The Building Blocks of the Body

getting started with

anatomy & physiology

In this part . . .

✔ Explore the basic building blocks and functions that make the parts of the body what they are. Dig into atoms, elements, chemical reactions, and metabolism.

✔ Crack open the cell to see what's happening at life's most elemental levels. Find out about the cell membrane, the nucleus, organelles, proteins, and the cell life cycle.

✔ Plunge into cell division, which has several phases: interphase, prophase, metaphase, anaphase, telophase, and cytokinesis. As you find out, sometimes things go wrong during this division.

✔ Use histology to build all of the body's tissues — epithelial, connective, muscle, and nerve — from the inside out.

Chapter 1

The Chemistry of Life

*W*e can hear your cries of alarm. You thought you were getting ready to learn about the knee bone connecting to the thigh bone. How in the heck does that involve (horrors!) *chemistry?* As much as you may not want to admit it, chemistry — particularly *organic chemistry,* the branch of the field that focuses on carbon-based molecules — is a crucial starting point for understanding how the human body works. When all is said and done, the universe boils down to two fundamental components: *matter,* which occupies space and has mass; and *energy,* the ability to do work or create change. In this chapter, we review the interactions between matter and energy to give you some insight into what you need to know to ace those early-term tests.

Building from Scratch: Atoms and Elements

All matter — be it solid, liquid, or gas — is composed of atoms. An *atom* is the smallest unit of matter capable of retaining the identity of an element during a chemical reaction. An *element* is a substance that can't be broken down into simpler substances by normal chemical reactions. There are 98 naturally occurring elements in nature and 20 (at last count) artificially created elements for a total of 118 known elements. However, additional spaces have yet to be filled in on the periodic chart of elements, which organizes all the elements by name, symbol, atomic weight, and atomic number. The key elements of interest to students of anatomy and physiology are

- ✔ **Hydrogen:** Symbol H
- ✔ **Oxygen:** Symbol O
- ✔ **Nitrogen:** Symbol N
- ✔ **Carbon:** Symbol C

HONC your horn for the four organic elements. These four elements make up 96 percent of all living material.

Atoms are made up of the subatomic particles *protons* and *neutrons,* which are in the atom's *nucleus,* and clouds of *electrons* orbiting the nucleus. The *atomic weight,* or *mass,* of an atom is the total number of protons and neutrons in its nucleus. The *atomic number* of an atom is its number of protons; conveniently, atoms that are electrically neutral have the same number of positive charges as negative charges. Opposite charges attract, so negatively charged electrons are attracted to positively charged protons. The attraction holds electrons in orbits outside the nucleus. The more protons there are in the nucleus, the stronger the atom's positive charge is and the more electrons it can attract.

Simplified models assume that electron particles orbit the nucleus at different energy levels, known as *shells* (see Figure 1-1). But quantum theory best describes information about electron-states using *wave functions,* or *orbitals.* Each orbital has a four-quantum-number address (characteristic energy, 3-D shape, orientation, and spin). The *principal quantum number* is the *shell.* The other numbers allow room to add distinct electrons. *Shells* are divided into *subshells,* and these are filled to completion in order of increasing energy.

- The first shell holds only two electrons, of opposite *spin (magnetism).*

- The second and third shells each hold eight electrons in four subshells and two spins.

- The fourth shell (which can be found in elements such as potassium, calcium, and iron) holds up to 18 electrons. Higher shells also exist.

Figure 1-1: Grouping electrons into shells or orbits.

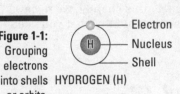

HYDROGEN (H) — Electron, Nucleus, Shell

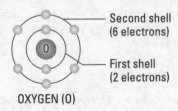

OXYGEN (O) — Second shell (6 electrons), First shell (2 electrons)

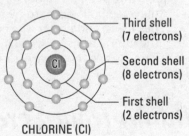

CHLORINE (Cl) — Third shell (7 electrons), Second shell (8 electrons), First shell (2 electrons)

© John Wiley & Sons, Inc.

Other key chemistry terms that you need to know are

- **Isotopes:** Atoms of an element that have a different number of neutrons and a different atomic weight than usual. In other words, isotopes are alternate forms of the same chemical element, so they always have the same number of protons as that element but a different number of neutrons.

- **Ions:** Because electrons are relatively far from the atomic nucleus, they are most susceptible to external fields. Atoms that have gained or lost electrons are transformed into ions. Getting an extra electron turns an atom into a negatively charged ion, or *anion,* whereas losing an electron creates a positively charged ion, or *cation.*

   To keep anions and cations straight, think like a compulsive dieter: Gaining is negative, and losing is positive.

- **Acid:** A substance that becomes ionized when placed in solution, producing positively charged hydrogen ions, H^+. An acid is considered a proton donor. (Remember, atoms always have the same number of electrons as protons. Ions are produced when an atom gains or loses electrons.) Stronger acids separate into larger numbers of H^+ ions in solution.

✔ **Base:** A substance that becomes ionized when placed in solution, producing negatively charged hydroxide ions, OH⁻. Bases are referred to as being more *alkaline* than acids and are known as proton acceptors. Stronger bases separate into larger numbers of OH⁻ ions in solution.

✔ **pH (potential of hydrogen):** A mathematical measure on a scale of 0 to 14 of the acidity or alkalinity of a substance. A solution is considered *neutral,* neither acid nor base, if its pH is exactly 7. (Pure water has a pH of 7.) A substance is *basic* if its pH is greater than 7 and *acidic* if its pH is less than 7. The strength of an acid or base is quantified by its absolute difference from that neutral number of 7. This number is large for a strong base and small for a weak base. Interestingly, skin is considered acidic because it has a pH around 5. Blood, on the other hand, is basic with a pH around 7.4.

Answer these practice questions about atoms and elements:

1. What are the four key elements that make up most living matter?

   a. Carbon, hydrogen, nitrogen, and phosphorus

   b. Oxygen, carbon, sulfur, and nitrogen

   c. Hydrogen, nitrogen, oxygen, and carbon

   d. Nitrogen, potassium, carbon, and oxygen

2. Among the subatomic particles in an atom, the two that have equal weight are

   a. neutrons and electrons.

   b. protons and neutrons.

   c. positrons and protons.

   d. neutrons and positrons.

3. For an atom with an atomic number of 19 and an atomic weight of 39, the total number of neutrons is

   a. 19.

   b. 20.

   c. 39.

   d. 58.

4. Element X has 14 electrons. How many electrons are in its outermost shell?

   a. 2

   b. 6

   c. 14

   d. 4

5. A substance that, in water, separates into a large number of hydroxide ions is

   a. a weak acid.

   b. a weak base.

   c. a strong acid.

   d. a strong base.

6. A hydroxide ion has an oxygen atom as well as

   a. a hydrogen atom.

   b. an extra electron.

   c. a hydrogen atom and an extra electron.

   d. a hydrogen atom and one less electron.

7.–12. Fill in the blanks to complete the following sentences:

Different isotopes of the same element have the same number of **7.** _____ and **8.** _____ but different numbers of **9.** _____. Isotopes also have different atomic **10.** _____. An atom that gains or loses an electron is called an **11.** _____. If an atom loses an electron, it carries a **12.** _____ charge.

Compounding Chemical Reactions

A *chemical reaction* is the result of a process that changes the number, the types, or the arrangement of atoms within a molecule. The substances that go through this process are called the *reactants*. The substances produced by the reaction are called the *products*. Chemical reactions are written in the form of an equation, with an arrow indicating the direction of the reaction. For instance: $A + B \rightarrow AB$. This equation translates to: Atom, ion, or molecule A plus atom, ion, or molecule B yields molecule AB.

In the following sections, we describe the chemical bonds that hold together molecules and the organic compounds created by chemical reactions.

Chemical bonds

Atoms tend to arrange themselves in the most stable patterns possible, which means that they have a tendency to complete or fill their outermost electron orbits. They join with other atoms to do just that. The force that holds atoms together in collections known as *molecules* is referred to as a *chemical bond*. There are two main types and some secondary types of chemical bonds:

   ✔ **Ionic bond:** This chemical bond (shown in Figure 1-2) involves a transfer of an electron, so one atom gains an electron while one atom loses an electron. One of the resulting ions carries a negative charge (anion), and the other ion carries a positive charge (cation). Because opposite charges attract, the atoms bond together to form a molecule.

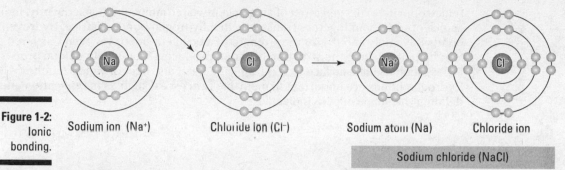

Figure 1-2:
Ionic
bonding.

Sodium ion (Na⁺)   Chloride ion (Cl⁻)   Sodium atom (Na)   Chloride ion

Sodium chloride (NaCl)

© John Wiley & Sons, Inc.

✔ **Covalent bond:** The most common bond in organic molecules, a covalent bond (shown in Figure 1-3) involves the sharing of electrons between two atoms. The pair of shared electrons forms a new orbit that extends around the nuclei of both atoms, producing a molecule. There are two secondary types of covalent bonds that are relevant to biology:

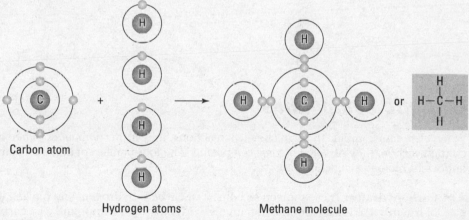

Carbon atom   or   H—C—H

Figure 1-3:
Covalent
bonding.

Hydrogen atoms   Methane molecule

© John Wiley & Sons, Inc.

- **Polar bond:** Two atoms connected by a covalent bond may exert different attractions for the electrons in the bond, producing an unevenly distributed charge. The result is known as a *polar bond* (shown in Figure 1-4), an intermediate case between ionic and covalent bonding, with one end of the molecule slightly negatively charged and the other end slightly positively charged. These slight imbalances in charge distribution are indicated in Figure 1-4 by lowercase delta symbols (δ) with a charge superscript (+ or −). Although the resulting molecule is neutral, at close distances the uneven charge distribution can be important. Water is an example of a polar molecule; the oxygen end has a slight positive charge whereas the hydrogen ends are slightly negative. Polarity explains why some substances dissolve readily in water and others do not.

- **Hydrogen bond:** Because they're polarized, two adjacent H_2O (water) molecules can form a linkage known as a *hydrogen bond* (see Figure 1-4), where the (electronegative) hydrogen atom of one H_2O molecule is electrostatically attracted to the (electropositive) oxygen atom of an adjacent water molecule. Consequently, molecules of water join together transiently in a hydrogen-bonded lattice. Hydrogen bonds have only about 1/20 the strength of a covalent bond, yet even this force is sufficient to affect the structure of water, producing many of its unique properties, such as high surface tension, specific heat, and heat of vaporization. Hydrogen bonds are important in many life processes, such as in replication and defining the shape of DNA molecules.

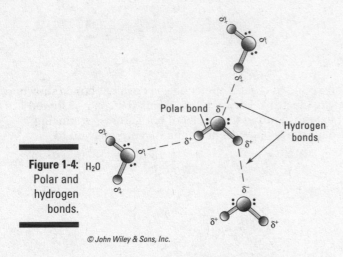

Figure 1-4: H_2O Polar and hydrogen bonds.

© John Wiley & Sons, Inc.

Organic compounds

When elements combine through chemical reactions, they form *compounds*. When compounds contain carbon, they're called *organic compounds*. The four families of organic compounds with important biological functions are

✔ **Carbohydrates:** These molecules consist of carbon, hydrogen, and oxygen in a ratio of roughly 1:2:1. If a test question involves identifying a compound as a carbohydrate, count the atoms and see if they fit that ratio. Carbohydrates are formed by the chemical reaction process of *condensation,* or *dehydration synthesis,* and broken apart by *hydrolysis,* the cleavage of a chemical by a reaction that adds water. There are several subcategories of carbohydrates:

  • *Monosaccharides,* also called *monomers* or *simple sugars,* are the building blocks of larger carbohydrate molecules and are a source of stored energy (see Figure 1-5). Key monomers include *glucose* (also known as blood sugar), *fructose,* and *galactose.* These three have the same numbers of carbon (6), hydrogen (12), and oxygen (6) atoms in each molecule — formally written as $C_6H_{12}O_6$ — but the bonding arrangements are different. Molecules with this kind of relationship are called *isomers.* Two important five-carbon monosaccharides (pentoses) are *ribose,* a component of ribonucleic acids (RNA), and *deoxyribose,* a component of deoxyribonucleic acids (DNA).

  • *Disaccharides,* or *dimers,* are sugars formed by the bonding of two monosaccharides, including *sucrose* (table sugar), *lactose,* and *maltose.*

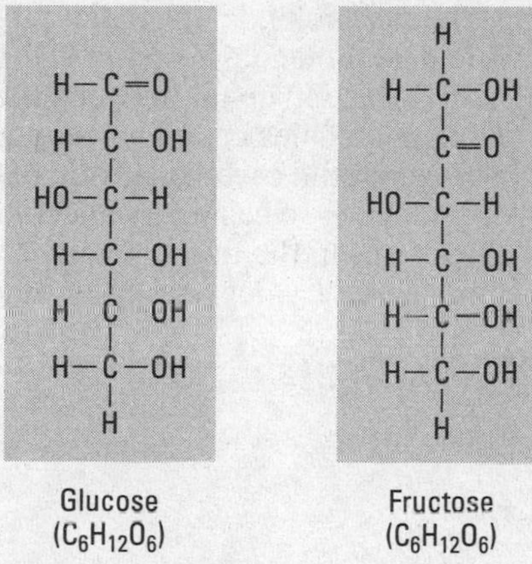

Figure 1-5: Monosaccharides.

Glucose ($C_6H_{12}O_6$)

Fructose ($C_6H_{12}O_6$)

© John Wiley & Sons, Inc.

- *Oligosaccharides* (from the Greek *oligo,* a few, and *sacchar,* sugar) contain three to nine simple sugars that serve many functions. They are found on plasma membranes of cells where they function in cell-to-cell recognition.

- *Polysaccharides,* or *polymers,* are formed when many monomers bond into long, chainlike molecules. *Glycogen* is the primary polymer in the body; it breaks down to individual monomers of glucose, an immediate source of energy for cells.

✔ **Lipids:** Commonly known as fats, these molecules contain carbon, hydrogen, and oxygen, and sometimes nitrogen and phosphorous. Insoluble in water because they contain a preponderance of nonpolar bonds, lipid molecules have six times more stored energy than carbohydrate molecules. Upon hydrolysis, however, most fats form glycerol and fatty acids. A fatty acid is a long, straight chain of carbon atoms with hydrogen atoms attached (see Figure 1-6). If the carbon chain has its full number of hydrogen atoms, the fatty acid is *saturated* (examples include butter and lard). If the carbon chain has less than its full number of hydrogen atoms, the fatty acid is *unsaturated* (examples include margarine and vegetable oils). All fatty acids contain a carboxyl or acid group, –COOH, at the end of the carbon chain. *Phospholipids,* as the name suggests, contain phosphorus and often nitrogen and form a bilayer in the cell membrane. *Steroids* are fat-soluble compounds such as vitamins A or D and hormones that often serve to regulate metabolic processes.

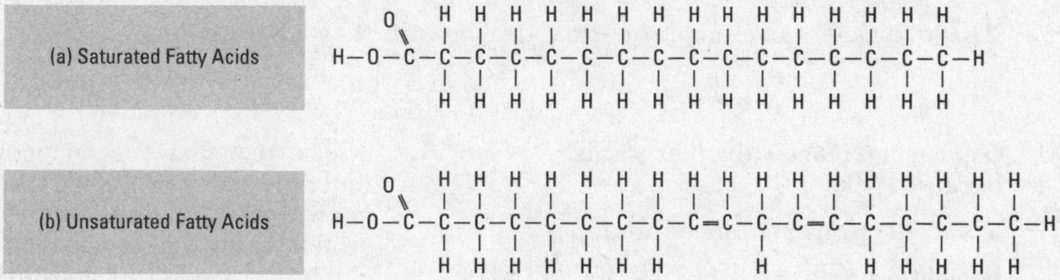

(a) Saturated Fatty Acids

(b) Unsaturated Fatty Acids

Figure 1-6: Fatty acids.

© John Wiley & Sons, Inc.

✔ **Proteins:** Among the largest molecules, proteins can reach molecular weights of some 40 million atomic units. Proteins always contain the four HONC elements — hydrogen, oxygen, nitrogen, and carbon — and sometimes contain phosphorus and sulfur. The human body builds protein molecules using 20 different kinds of smaller molecules called *amino acids* (see Figure 1-7). Each amino acid molecule is composed of an amino group, $-NH_2$, and a carboxyl group, $-COOH$, with a carbon chain between them. Amino acids link together by *peptide bonds* to form long molecules called *polypeptides,* which then assemble into proteins. These bonds form when the carboxyl group of one molecule reacts with the amino group of another molecule, releasing a molecule of water (*dehydration synthesis reaction*). Examples of proteins in the body include *antibodies, hemoglobin* (the red pigment in red blood cells), and *enzymes* (catalysts that accelerate reactions in the body).

amino acid   amino acid

Figure 1-7: Amino acids in a protein molecule.

protein molecule

© John Wiley & Sons, Inc.

✔ **Nucleic acids:** These long molecules, found primarily in the cell's nucleus, act as the body's genetic blueprint. They're comprised of smaller building blocks called *nucleotides.* Each nucleotide, in turn, is composed of a five-carbon sugar (*deoxyribose* or *ribose),* a phosphate group, and a nitrogenous base. The nitrogenous bases in DNA (deoxyribonucleic acid) are *adenine, thymine, cytosine,* and *guanine;* they always pair off A-T and C-G. In RNA (ribonucleic acid), which occurs in a single strand, thymine is replaced by *uracil,* so the nucleotides pair off A-U and C-G. In 1953, James Watson and Francis Crick published their discovery of the three-dimensional structure of DNA — a polymer that looks like a ladder twisted into a coil. They called this structure the *double-stranded helix* (see Figure 1-8).

The following is an example question dealing with chemical reactions:

Q. Oxygen can react with other atoms because it has

a. two electrons in its inner orbit.

b. eight protons.

c. an incomplete outer electron orbit.

d. eight neutrons.

A. The correct answer is an incomplete outer electron orbit. Even if you don't know the first thing about oxygen, remembering that atoms tend toward stability answers this question for you.

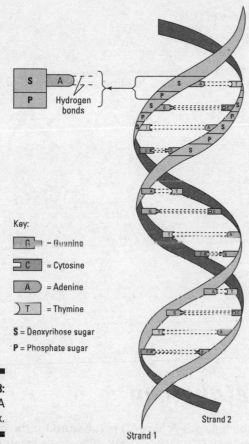

Key:

G = Guanine

C = Cytosine

A = Adenine

T = Thymine

S = Deoxyribose sugar

P = Phosphate sugar

Figure 1-8:
The DNA
double helix.

Strand 1

Strand 2

© John Wiley & Sons, Inc.

13. Covalent bonds are a result of

a. completing an inner electron orbit.

b. sharing one or more electrons between atoms.

c. deleting neutrons from both atoms.

d. reactions between protons.

14. The formation of chemical bonds is based on the tendency of an atom to

a. move protons into vacant electron orbit spaces.

b. fill its outermost energy level.

c. radiate excess neutrons.

d. pick up free protons.

15. Which of the following statements is *not* true of DNA?

   a. DNA is found in the nucleus of the cell.

   b. DNA can replicate itself.

   c. DNA contains the nitrogenous bases adenine, thymine, guanine, cytosine, and uracil.

   d. DNA forms a double-helix molecule.

16. Polysaccharides

   a. can be reduced to fatty acids.

   b. contain nitrogen and phosphorus.

   c. are complex carbohydrates.

   d. contain adenine and uracil.

17. Amino acids

   a. help reduce carbohydrates.

   b. are the building blocks of proteins.

   c. modulate the production of lipids.

   d. control nucleic acids.

Cycling through Life: Metabolism

Metabolism (from the Greek *metabole,* which means "change") is the word for the myriad chemical reactions that happen in the body, particularly as they relate to generating, storing, and expending energy. All metabolic reactions are either *catabolic* or *anabolic*.

   ✔ **Catabolic reactions** break down food molecules to release energy (memory tip: it can be *cata*strophic when things break down).

   ✔ **Anabolic reactions** require a source of energy to build up compounds that the body needs.

The chemical alteration of molecules in the cell is referred to as *cellular metabolism*. *Enzymes* can be used as catalysts, accelerating chemical reactions without being changed by the reactions. The molecules that enzymes react with are called *substrates*.

Adenosine triphosphate (ATP) is a molecule that stores energy in a cell until the cell needs it. As the *tri–* prefix implies, a single molecule of ATP is composed of three phosphate groups attached to a nitrogenous base of adenine. ATP's energy is stored in high-energy bonds that attach the second and third phosphate groups. (The high-energy bond is symbolized by a wavy line.) When a cell needs energy, it removes one or two of these phosphate groups, releasing energy and converting ATP into either the two-phosphate molecule *adenosine diphosphate* (ADP) or the one-phosphate molecule *adenosine monophosphate* (AMP). (You can see ADP and ATP molecules in Figure 1-9.) Later, through additional metabolic reactions, the second and third phosphate groups containing energy are reattached to adenosine, reforming an ATP molecule until energy is needed again.

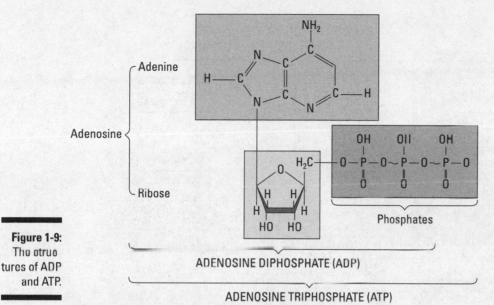

Figure 1-9:
The structures of ADP and ATP.

Oxidation-reduction (redox) reactions are an important pair of reactions that occur in carbohydrate, lipid, and protein metabolism (see Figure 1-10). When a substance is *oxidized,* it loses electrons. When a substance is *reduced,* it gains electrons. Oxidation and reduction occur together, so whenever one substance is oxidized, another is reduced. The body uses this chemical-reaction pairing to transport energy in a process known as the respiratory chain, or the *electron transport chain.*

Carbohydrate metabolism involves a series of cellular respiration reactions, which are illustrated in Figure 1-11. All food carbohydrates are eventually broken down into glucose; therefore, carbohydrate metabolism is really glucose metabolism. Glucose metabolism produces energy that is then stored in ATP molecules. The oxidation process in which energy is released from molecules, such as glucose, and transferred to other molecules is called *cellular respiration.* It occurs in every cell in the body, and it is the cell's source of energy. The complete oxidation of one molecule of glucose will produce 38 molecules of ATP. It occurs in three stages: *glycolysis,* the *Krebs cycle,* and the *electron transport chain:*

1. **Glycolysis**

   From the Greek *glyco* (sugar) and *lysis* (breakdown), this is the first stage of both *aerobic* (with oxygen) and *anaerobic* (without oxygen) respiration. Using energy from two molecules of ATP and two molecules of NAD$^+$ *(nicotinamide adenine di-nucleotide),* glycolysis uses a process called *phosphorylation* to convert a molecule of six-carbon glucose — the smallest molecule that the digestive system can produce during the breakdown of a carbohydrate — into two molecules of three-carbon *pyruvic acid* or *pyruvate,* as well as four ATP molecules and two molecules of NADH *(nicotinamide adenine dinucleotide).* Taking place in the cell's cytoplasm (see Chapter 2), glycolysis doesn't require oxygen to occur. The pyruvate and NADH move into the cell's *mitochondria* (detailed in Chapter 2), where an aerobic (with oxygen) process converts them into ATP.

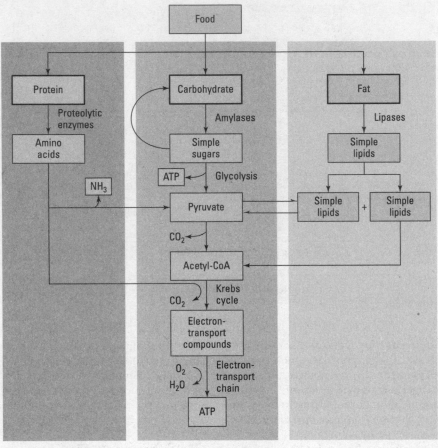

Figure 1-10:
Protein, carbohydrate, and fat metabolism.

© John Wiley & Sons, Inc.

2. Krebs cycle

Also known as the *tricarboxylic acid cycle* or *citric acid cycle,* this series of energy-producing chemical reactions begins in the mitochondria after pyruvate arrives from glycolysis. Before the Krebs cycle can begin, the pyruvate loses a carbon dioxide group to form *acetyl coenzyme A* (acetyl CoA). Then acetyl CoA combines with a four-carbon molecule (*oxaloacetic acid,* or OAA) to form a six-carbon citric acid molecule that then enters the Krebs cycle. The CoA is released intact to bind with another acetyl group. During the conversion, two carbon atoms are lost as carbon dioxide, and energy is released. One ATP molecule is produced each time an acetyl CoA molecule is split. The cycle goes through eight steps, rearranging the atoms of citric acid to produce different intermediate molecules called *keto acids.* The acetic acid is broken apart by carbon (or *decarboxylated*) and oxidized, generating three molecules of NADH, one molecule of FADH2 (flavin adenine dinucleotide), and one molecule of ATP. The energy can be transported to the electron transport chain and used to produce more molecules of ATP. OAA is regenerated to get the next cycle going, and carbon dioxide produced during this cycle is exhaled from the lungs.

3. Electron transport chain

The electron transport chain is a series of energy compounds attached to the inner mitochondrial membrane. The electron molecules in the chain are called *cytochromes.*

These electron-transferring proteins contain a heme, or iron, group. Hydrogen from oxidized food sources attaches to coenzymes that in turn combine with molecular oxygen. The energy released during these reactions is used to attach inorganic phosphate groups to ADP and form ATP molecules.

Pairs of electrons transferred to NAD^+ go through the electron transport process and produce three molecules of ATP by oxidative phosphorylation. Pairs of electrons transferred to FAD enter the electron transport after the first phosphorylation and yield only two molecules of ATP. Oxidative phosphorylation is important because it makes energy available in a form the cells can use.

At the end of the chain, two positively charged hydrogen molecules combine with two electrons and an atom of oxygen to form water. The final molecule to which electrons are passed is oxygen. Electrons are transferred from one molecule to the next, producing ATP molecules.

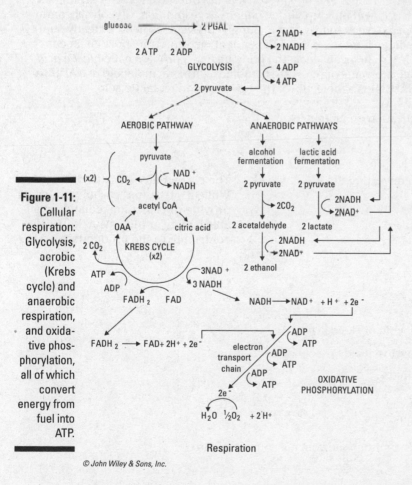

Figure 1-11: Cellular respiration: Glycolysis, aerobic (Krebs cycle) and anaerobic respiration, and oxidative phosphorylation, all of which convert energy from fuel into ATP.

© John Wiley & Sons, Inc.

Lipid metabolism (refer to Figure 1-10) only requires portions of the processes involved in carbohydrate metabolism. Lipids contain about 99 percent of the body's stored energy and can be digested at mealtime, but as people who complain about fats going "straight to their hips" can attest, lipids are more inclined to be stored in *adipose tissue* — the stuff generally identified with body fat. When the body is ready to metabolize lipids, a series of catabolic reactions breaks apart two carbon atoms from the end of a fatty acid chain to form acetyl CoA, which then enters the Krebs cycle to produce ATP. Those reactions continue to strip two carbon atoms at a time until the entire fatty acid chain is converted into acetyl CoA molecules.

Protein metabolism (refer to Figure 1-10) focuses on producing the amino acids needed for synthesis of protein molecules within the body. But in addition to the energy released into the electron transport chain during protein metabolism, the process also produces byproducts, such as ammonia and keto acid. Energy is released entering the electron transport chain. The liver converts the ammonia into urea, which the blood carries to the kidneys for elimination. The keto acid enters the Krebs cycle and is converted into pyruvic acids to produce ATP.

One last thing: That severe soreness and fatigue you feel in your muscles after strenuous exercise is the result of lactic acid buildup during *anaerobic respiration*. Glycolysis continues because it doesn't need oxygen to take place. But glycolysis does need a steady supply of NAD^+, which usually comes from the oxygen-dependent electron transport chain converting NADH back into NAD^+. In its absence, the body begins a process called *lactic acid fermentation,* in which one molecule of pyruvate combines with one molecule of NADH to produce a molecule of NAD^+ plus a molecule of the toxic byproduct lactic acid.

Check out an example question on metabolism:

Q. Cells obtain ATP by converting the energy in

    a. carbohydrates.

    b. proteins.

    c. lipids.

    d. all of these.

A. The correct answer is all of these. While it's true that carbohydrates provide the most immediately available energy, proteins and lipids also contribute to the production of ATP.

18. During glycolysis, a molecule of glucose is

    a. broken down to two pyruvic acids.

    b. broken down to two lipids.

    c. converted into energy.

    d. used to oxidate proteins.

19. After pyruvic acid enters a mitochondrion, it's converted into

    a. glucose.

    b. acetyl CoA.

    c. water.

    d. protein.

20. A molecule of glucose can be converted into how many ATP molecules?

   a. 2

   b. 3

   c. 38

   d. 45

21. What is the part of metabolism that involves creating compounds the body needs?

   a. A catabolic reaction

   b. Cellular respiration

   c. An anabolic reaction

   d. Oxidation

22. Anaerobic metabolic processes that don't require oxygen can lead to

   a. buildup of lactic acid.

   b. excess protein in the bloodstream.

   c. mitochondrial exhaustion.

   d. carbon dioxination.

23. Which two respiration processes take place in the cell's mitochondria?

   a. Glycolysis and the Krebs cycle

   b. Glycolysis and the electron transport chain

   c. The Krebs cycle and the electron transport chain

   d. The Krebs cycle and anaerobic respiration

24. Coal is to electricity as glucose is to

   a. ATP.

   b. pyruvate.

   c. hydrogen.

   d. glycolysis.

25. What are the primary products of protein metabolism?

   a. ATP molecules

   b. Amino acids

   c. Lipids

   d. Carbon dioxide molecules

26. A lactic acid chain is converted to acetyl CoA, which then produces ATP in

   a. glycolysis.

   b. lactic acid fermentation.

   c. exercise.

   d. the Krebs cycle.

Answers to Questions on Life's Chemistry

The following are answers to the practice questions presented in this chapter.

1 The four key elements that make up most living matter are **c. hydrogen, nitrogen, oxygen, and carbon.** We arranged them so that they spell HNOC instead of HONC, but you get the idea, right?

2 Among the subatomic particles in an atom, the two that have equal weight are **b. protons and neutrons.** That's why you add them together to determine atomic weight, or mass.

3 For an atom with an atomic number of 19 and an atomic weight of 39, the total number of neutrons is **b. 20.** The atomic number of 19 is the same as the number of protons. The atomic weight of 39 tells you the number of protons plus the number of neutrons: $39 - 19 = 20$.

4 Element X has 14 electrons. How many electrons are in its outermost shell? **d. 4.** The first orbit has the maximum two electrons, and the second orbit has the maximum eight electrons. That makes ten electrons in the first two orbits, leaving only four for the third, outermost orbit.

5 A substance that, in water, separates into a large number of hydroxide ions is **d. a strong base.** The more hydroxide ions there are, the stronger the base is.

6 A hydroxide ion has an oxygen atom as well as **c. a hydrogen atom and an extra electron.** The first few letters of the word "hydroxide" are a dead giveaway that there's a hydrogen atom in there; plus, hydroxide ions are negatively charged, which calls for that extra electron.

7–12 Different isotopes of the same element have the same number of **7. electrons/protons** and **8. protons/electrons** but different numbers of **9. neutrons.** Isotopes also have different atomic **10. weights.** An atom that gains or loses an electron is called an **11. ion.** If an atom loses an electron, it carries a **12. positive** charge.

13 Covalent bonds are a result of **b. sharing one or more electrons between atoms.** If the atoms had gained or lost electrons, it would be an ionic bond, but here they're sharing — valiantly cohabiting, if you will.

14 The formation of chemical bonds is based on the tendency of an atom to **b. fill its outermost energy level.** This is true whether an atom fills its outer shell by sharing, gaining, or losing electrons.

15 Which of the following statements is _not_ true of DNA? **c. DNA contains the nitrogenous bases adenine, thymine, guanine, cytosine, and uracil.** This statement is false because only RNA contains uracil.

16 Polysaccharides **c. are complex carbohydrates.** The root _poly–_ means "many," which you can interpret as "complex." The root _mono–_ means "one," which you can interpret as "simple."

17 Amino acids **b. are the building blocks of proteins.** Being such large molecules, proteins need to be built from complex molecules to begin with.

18 During glycolysis, a molecule of glucose is **a. broken down to two pyruvic acids.** Remember that glucose must become pyruvic acid before it enters the Krebs cycle.

19 Pyruvic acid enters a mitochondrion and is converted into **b. acetyl CoA.** Don't forget that the Krebs cycle, during which pyruvate is broken down, occurs in the mitochondrion.

20 A molecule of glucose can be converted into how many ATP molecules? **c. 38.** Two net molecules of ATP come from glycolysis, two molecules come from the Krebs cycle, and the electron transport chain churns out 34.

21 The part of metabolism that involves creating compounds the body needs is called **c. an anabolic reaction.** Breaking things down is a catabolic reaction, but building them up is an anabolic reaction.

22 Anaerobic metabolic processes that don't require oxygen can lead to **a. buildup of lactic acid.** Recall that during aerobic exercise, you're trying to circulate oxygen to your muscles. During anaerobic activities, the buildup of lactic acid makes your muscles sore and fatigued.

23 Which two respiration processes take place in the cell's mitochondria? **c. The Krebs cycle and the electron transport chain.** The other answers are incorrect because glycolysis takes place in the cytoplasm, and anaerobic respiration isn't one of the three cellular respiration processes.

24 Coal is to electricity as glucose is to **a. ATP.** Just as you can't power a lamp with a lump of coal, cells can't use glucose directly. You need to turn the coal into electricity, and cells need to turn the glucose into ATP.

25 The primary products of protein metabolism are **b. amino acids.** Although some ATP comes from metabolizing proteins, the body primarily needs to get amino acids from any protein that's consumed.

26 A lactic acid chain is converted to acetyl CoA, which then produces ATP in **d. the Krebs cycle.** That's the only process that can use the acetyl CoA supplied by lipids.

Chapter 2

The Cell: Life's Basic Building Block

Cytology, from the Greek word *cyto,* which means "cell," is the study of cells. Every living thing has cells, but not all living things have the same kinds of cells. *Eukaryotes* like humans (and all other organisms besides bacteria and viruses) have *eukaryotic cells,* each of which has a defined *nucleus* that controls and directs the cell's activities, and *cytosol,* fluid material found in the gel-like *cytoplasm* that fills most of the cell. Plant cells have fibrous cell walls; animal cells have a semipermeable cell membrane called a *plasma membrane* or the *plasmalemma.* Because human cells don't have cell walls, they look like gel-filled sacs with nuclei and tiny parts called *organelles* nestled inside when viewed through an electron microscope.

In this chapter, we help you sort out what makes up a cell, what all those tiny parts do, and how cells act as protein-manufacturing plants to support life's activities. We then take a quick look at an individual cell's life cycle.

Gaining Admission: The Cell Membrane

Think of it as a gatekeeper, guardian, or border guard. Despite being only 6 to 10 nanometers thick and visible only through an electron microscope, the cell membrane keeps the cell's cytoplasm in place and lets only select materials enter and depart the cell as needed. This semipermeability, or *selective permeability,* is a result of a double layer (bilayer) of *phospholipid* molecules interspersed with protein molecules. The outer surface of each layer is made up of tightly packed *hydrophilic* (or water-loving) *polar heads.* Inside, between the two layers, you find *hydrophobic* (or water-fearing) *nonpolar tails* consisting of fatty acid chains. (See Figure 2-1.)

Cholesterol molecules between the phospholipid molecules give the otherwise elastic membrane stability and make it less permeable to water-soluble substances. Both cytoplasm and the *matrix,* the material in which cells lie, are primarily water. The polar heads electrostatically attract polarized water molecules while the nonpolar tails lie between the layers, shielded from water and creating a dry middle layer. The membrane's interior is made up of oily fatty acid molecules that are electrostatically symmetric, or *nonpolarized*. Lipid-soluble molecules can pass through this layer, but water-soluble molecules such as amino acids, sugars, and proteins cannot, instead moving through the membrane via transport channels made by embedded *channel proteins*. Because phospholipids have both polar and nonpolar regions, they're also called *amphipathic molecules*.

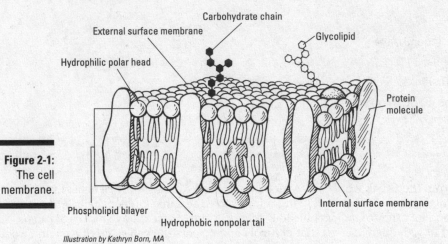

Figure 2-1:
The cell
membrane.

Illustration by Kathryn Born, MA

The cell membrane is designed to hold the cell together and to isolate it as a distinct functional unit of protoplasm. Although it can spontaneously repair minor tears, severe damage to the membrane will cause the cell to disintegrate. The membrane is picky about which molecules it lets in or out. It allows movement across its barrier by *diffusion, osmosis,* or *active transport*.

Diffusion

Diffusion is a natural phenomenon with observable effects like *Brownian motion*. Molecules or other particles spontaneously spread, or migrate, from areas of higher concentration to areas of lower concentration until equilibrium occurs. At equilibrium, diffusion continues, but the net flow balances except for random fluctuations. This occurs because all molecules possess kinetic energy of random motion. They move at high speeds, colliding with one another, changing directions, and moving away from areas of greater concentration to areas of lower concentration. The diffusion rate depends on the mass and temperature of the molecule; lighter and warmer molecules move faster.

Diffusion is one form of *passive transport* that doesn't require the expenditure of cellular energy. A molecule can diffuse passively through the cell membrane if it's lipid-soluble, uncharged, and very small, or if a carrier molecule can assist it. The unassisted diffusion of very small or lipid-soluble particles is called *simple diffusion*. The assisted process is known as *facilitated diffusion*. The cell membrane allows nonpolar molecules (those that don't readily bond with water) to flow from an area where they're highly concentrated to an area where they're less concentrated. Embedded in the membrane are transmembrane protein molecules called *channel proteins* that traverse from the outer layer to the inner layer and create diffusion-friendly openings for molecules to move through.

Osmosis

Osmosis is a form of passive transport that's similar to diffusion and involves a solvent moving through a selectively permeable or semipermeable membrane from an area of higher concentration to an area of lower concentration. Solutions are composed of two parts: a solvent and a solute.

- The *solvent* is the liquid in which a substance is dissolved; water is called the *universal solvent* because more materials dissolve in it than in any other liquid.

- A *solute* is the substance dissolved in the solvent.

Typically, a cell contains a roughly 1 percent saline solution — in other words, 1 percent salt (solute) and 99 percent water (solvent). Water is a polar molecule that will not pass through the lipid bilayer; however, it's small enough to move through the pores — formed by protein molecules — of most cell membranes. Osmosis occurs when there's a difference in molecular concentration of water on the two sides of the membrane. The membrane allows the solvent (water) to move through but keeps out the solute (the particles dissolved in the water).

Transport by osmosis is affected by the concentration of solute (the number of particles) in the water. One molecule or one ion of solute displaces one molecule of water. *Osmolarity* is the term used to describe the concentration of solute particles per liter. As water diffuses into a cell, hydrostatic pressure builds within the cell. Eventually, the pressure within the cell becomes equal to, and is balanced by, the osmotic pressure outside.

- An *isotonic* solution has the same concentration of solute and solvent as found inside a cell, so a cell placed in isotonic solution — typically 1 percent saline solution for humans — experiences equal flow of water into and out of the cell, maintaining equilibrium.

- A *hypotonic* solution has less solute and higher water potential than inside the cell. An example is 100 percent distilled water, which has less solute than what is inside the cell. Therefore, if a human cell is placed in a hypotonic solution, molecules diffuse down the concentration gradient until the cell's membrane bursts.

- A *hypertonic* solution has more solute and lower water potential than inside the cell. So the membrane of a human cell placed in 10 percent saline solution (10 percent salt and 90 percent water) would let water flow out of the cell (from the higher concentration inside to the lower concentration outside), therefore shrinking it.

Active transport

Active transport occurs across a semipermeable membrane against the normal concentration gradient, moving from the area of lower concentration to the area of higher concentration and requiring an expenditure of energy released from an ATP molecule (as discussed in Chapter 1). Embedded with the hydrophilic heads in the outer layer of the membrane are transmembrane protein molecules able to detect and move compounds through the membrane. These *carrier* or *transport* proteins interact with the *passenger* molecules and use the ATP-supplied energy to move them against the gradient. The carrier molecules combine with the transport molecules — most importantly amino acids and ions — to pump them against their concentration gradients.

Active transport lets cells obtain nutrients that can't pass through the membrane by other means. In addition, there are secondary active transport processes that are similar to diffusion but instead use imbalances in electrostatic forces to move molecules across the membrane.

1.–3. Fill in the blanks to complete the following sentences:

The lipid bilayer structure of the cell membrane is made possible because phospholipid molecules contain two distinct regions: The **1.** _____ region is attracted to water, and the **2.** _____ region is repelled by water. Because it has both polar and nonpolar regions, a phospholipid is classified as a(n) **3.** _____ molecule.

4. What is osmosis?

   a. The diffusion of water molecules through the body of a cell

   b. The filtration of lipids from water molecules

   c. The movement of water molecules through a semipermeable membrane

   d. The repulsion of water by a cell's membrane

5. What is a hypotonic solution?

   a. A solution that has a lower concentration of water than exists in the cell

   b. A solution that has a greater concentration of water than exists in the cell

   c. A solution that has the same concentration of water as exists in the cell

   d. A solution that constantly varies in concentration when compared to what exists in the cell

6. Injecting a large quantity of distilled water into a human's veins would cause many red blood cells to

   a. swell and burst.

   b. shrink.

   c. carry more oxygen.

   d. aggregate.

7. The cell membrane does *not* function

   a. in selective transport of materials into and out of the cell.

   b. as a barrier protecting the cell.

   c. in the production of energy.

   d. as containment for the cytoplasm.

Aiming for the Nucleus

The cell nucleus is the largest cellular organelle and the first to be discovered by scientists. On average, it accounts for about 10 percent of the total volume of the cell, and it holds a complete set of genes.

The outermost part of this organelle is the *nuclear envelope,* which is composed of a double-membrane barrier, each membrane of which is made up of a phospholipid bilayer. Between the two membranes is a fluid-filled space called the *perinuclear cisterna.* The two layers fuse to form a selectively permeable barrier, but large *pores* allow relatively free movement of molecules and ions, including large protein molecules. Intermediate filaments lining the surface of the nuclear envelope make up the *nuclear lamina,* which functions in the disassembly and reassembly of the nuclear membrane during mitosis and binds the membrane to the endoplasmic reticulum. The nucleus also contains *nucleoplasm,* a clear viscous material that forms the matrix in which the organelles of the nucleus are embedded.

DNA is packaged inside the nucleus in linear structures called *chromatin,* or *chromatin networks.* During cell division (see Chapter 3), the chromatin contracts and becomes coiled, forming *chromosomes.* Chromosomes contain a DNA molecule encoded with the genetic information needed to direct the cell's activities. The most prominent subnuclear body is the *nucleolus,* a small spherical body that stores RNA molecules and produces *ribosomes,* which are exported to the cytoplasm where they translate *messenger RNA* (mRNA).

The following is an example question about the nucleus:

Q. The only cellular organelle found within the nucleus is called a(n)

   a. lamina.

   b. envelope.

   c. nucleolus.

   d. chromosome.

A. The correct answer is nucleolus. The other options aren't organelles. True, chromosomes are found inside the cell's nucleus, but they don't serve the cell as an organ might serve the body. Remember: The suffix "*–elle*" is a diminutive that makes the word *organelle* translate into "little organ."

8. What is a perinuclear cisterna?

   a. A fluid-filled space within the nuclear envelope

   b. A molecular machine that links amino acids

   c. A cellular "power plant" that generates ATP

   d. A protein packaging system within the cell

9. In the nucleus of a cell that isn't dividing, DNA

   a. moves about freely.

   b. is packaged within chromatin threads.

   c. regulates the activity of mitochondria.

   d. replenishes itself.

10. The nucleolus

    a. packages DNA.

    b. enables large molecule transport.

    c. forms a membrane around the nucleus.

    d. assembles ribosomes.

Looking Inside: Organelles and Their Functions

Molecules that pass muster with the cell membrane enter the *cytoplasm,* a mixture of macro-molecules (such as proteins and RNA), small organic molecules (such as glucose), ions, and water. Because of the various materials in the cytoplasm, it's a *colloid,* or mixture of phases, that alternates from a *sol* (a liquid colloid with solid suspended in it) to a *gel* (a colloid in which the dispersed phase combines with the medium to form a semisolid material). The fluid part of the cytoplasm, called the *cytosol,* has a differing consistency based on changes in temperature, molecular concentrations, pH, pressure, and agitation.

Within the cytoplasm lies a network of fibrous proteins collectively referred to as the *cyto-skeleton*. It's not rigid or permanent but changing and shifting according to the activity of the cell. The cytoskeleton maintains the cell's shape, enables it to move, anchors its organelles, and directs the flow of the cytoplasm. The fibrous proteins that make up the cytoskeleton include the following:

- *Microfilaments,* rodlike structures about 5 to 8 nanometers wide that consist of a stacked protein called *actin,* the most abundant protein in eukaryotic cells. They provide structural support and have a role in cell and organelle movement as well as in cell division.

- *Intermediate filaments,* the strongest and most stable part of the cytoskeleton. They average about 10 nanometers wide and consist of interlocking proteins, including *keratin,* that chiefly are involved in maintaining cell integrity and resisting pulling forces on the cell.

- Hollow *microtubules* about 25 nanometers in diameter and 25 to 50 microns long made of the protein *tubulin* that grow with one end embedded in the *centrosome* near the cell's nucleus. Like microfilaments, these components of cilia, flagella, and centrioles provide structural support, can shorten and elongate, and have a role in cell and organelle movement as well as in cell division.

Organelles, literally translated as "little organs," are nestled inside the cytoplasm (except for the two organelles that move, *cilia* and *flagellum,* which are found on the cell's exterior). Each organelle has different responsibilities for producing materials used elsewhere in the cell or body. Here are the key organelles and what they do:

- **Centrosome:** Microtubules sprout from this structure, which is located next to the nucleus and is composed of two *centrioles* — arrays of microtubules — that function in separating genetic material during cell division.

✔ **Cilia:** These are short, hairlike cytoplasmic projections on the external surface of the cell. In multicellular animals, including humans, cilia move materials over the surface of the cell. In some single-celled organisms, they're used for locomotion.

✔ **Endoplasmic reticulum (ER):** This organelle makes direct contact with the cell nucleus and functions in the transport of materials such as proteins and RNA molecules. Composed of membrane-bound canals and cavities that extend from the nuclear membrane to the cell membrane, the ER is the site of lipid and protein synthesis. The two types of ER are *rough,* which is dotted with ribosomes on the outer surface, and *smooth,* which has no ribosomes on the surface. The rough ER manufactures membranes and secretory proteins in collaboration with the ribosomes. The smooth ER has a wide range of functions, including carbohydrate and lipid synthesis.

✔ **Flagellum:** This whiplike cytoplasmic projection lies on the cell's exterior surface. Found in humans primarily on sperm cells, it's used for locomotion.

✔ **Golgi apparatus (or body):** This organelle consists of a stack of flattened sacs with membranes. Transport vesicles connect the ER with the Golgi apparatus, and secretory vesicles link the Golgi apparatus with the cell membrane. Located near the nucleus, it functions in the storage, modification, and packaging of proteins for secretion to various destinations within the cell.

✔ **Lysosome:** A tiny, membranous sac containing acids and digestive enzymes, the lysosome breaks down large food molecules such as proteins, carbohydrates, and nucleic acids into materials that the cell can use. It destroys foreign particles, including bacteria and viruses, and helps to remove nonfunctioning structures from the cell.

✔ **Mitochondrion:** Called the powerhouse of the cell, this usually rod-shaped organelle consists of two membranes — a smooth outer membrane and an invaginated (folded inward) inner membrane that divides the organelle into compartments. The inward-folding crevices of the inner membrane are called *cristae.* (See Figure 2-2.) The mitochondrion provides critical functions in cell respiration, including oxidizing (breaking down) food molecules and releasing energy that is stored in ATP molecules in the mitochondrion. This energy is used to accelerate chemical reactions in the cell, which we cover in Chapter 1.

✔ **Ribosomes:** These roughly 25-nanometer structures may be found along the endoplasmic reticulum or floating free in the cytoplasm. Composed of 60 percent RNA and 40 percent protein, they translate the genetic information on messenger RNA molecules to *synthesize,* or produce, a protein molecule.

✔ **Vacuoles:** More commonly found in plant cells, these open spaces in the cytoplasm sometimes carry materials to the cell membrane for discharge to the outside of the cell. In animal cells, food vacuoles are membranous sacs formed when food masses are pinched off from the cell membrane and passed into the cytoplasm of the cell. This process, called *endocytosis* (from the Greek words meaning "within the cell"), requires energy to move large masses of material into the cell. (*Exocytosis,* by contrast, is an energy-consuming process during which a cell directs the contents of secretory vesicles out of the cell membrane and into the extracellular space.) Vacuoles help to remove structural debris, isolate harmful materials, and export unwanted substances from the cell.

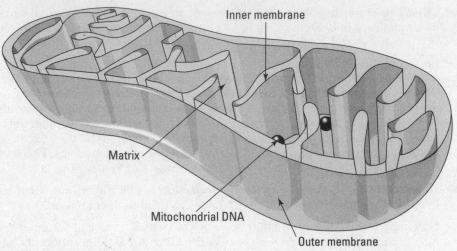

Inner membrane

Matrix

Mitochondrial DNA

Outer membrane

Figure 2-2:
A mighty
mitochon-
drion.

© John Wiley & Sons, Inc.

Answer these practice questions about cell organelles:

11. What is the function of a mitochondrion?

  a. To package and process proteins, lipids, and other macromolecules

  b. To produce energy through aerobic respiration

  c. To keep the cell clean and recycle materials within it

  d. To produce and export hormones

12. Actin, which is the most abundant protein in human cells,

  a. has filaments that contribute to the cytoskeleton.

  b. metabolizes lipids for cell energy.

  c. prevents binding of microtubules.

  d. synthesizes hormones.

13. Which organelle gets to take out the cellular trash?

  a. Golgi apparatus

  b. Ribosome

  c. Vacuole

  d. Endoplasmic reticulum

14. Which organelle translates genetic information to direct protein synthesis?

  a. The ribosome

  b. The lysosome

  c. The centriole

  d. The vesicle

15. Which organelle has ribosomes attached to it?

  a. Smooth (agranular) endoplasmic reticulum

  b. Golgi apparatus

  c. Rough (granular) endoplasmic reticulum

  d. Lysosome

16. Which organelle contains secretory materials?

  a. Golgi apparatus

  b. Ribosome

  c. Lysosome

  d. Endoplasmic reticulum

17. Which of the following can change the consistency of cytoplasm?

  a. Changes in acidity or alkalinity

  b. Temperature

  c. Pressure

  d. All of the above

18. Organelles don't just float freely within a cell; some are found inside the nucleus. Which of these can be found within the nucleus?

  a. Mitochondria

  b. Lysosomes

  c. Chromatin network

  d. Ribosomes

19.–32. Use the terms that follow to identify the cell structures and organelles shown in Figure 2-3.

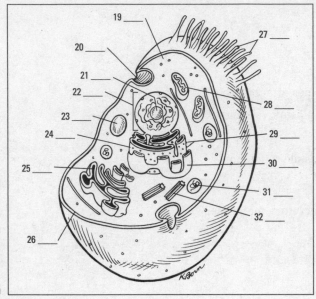

Figure 2-3:
A cutaway view of an animal cell and its organelles.

Illustration by Kathryn Born, MA

a. Centriole

b. Cilia

c. Cytoplasm

d. Golgi apparatus

e. Lysosome

f. Mitochondrion

g. Nucleolus

h. Nucleus

i. Plasma (cell) membrane

j. Ribosomes

k. Rough endoplasmic reticulum

l. Smooth endoplasmic reticulum

m. Vacuole

n. Vesicle formation

33.–37. Match the organelles with their descriptions.

33. _____ Mitochondrion

34. _____ Nucleolus

35. _____ Flagellum

36. _____ Cytoplasm

37. _____ Lysosomes

a. Long, whiplike organelle for locomotion

b. Fluidlike interior of the cell that may become a semisolid, or colloid

c. Membranous sacs containing digestive enzymes

d. Powerhouse of the cell

e. Stores RNA in the nucleus

Putting Together New Proteins

Proteins are essential building blocks for all living systems, which helps explain why the word is derived from the Greek term *proteios*, meaning "holding first place." Cells use proteins to perform a variety of functions, including providing structural support and catalyzing reactions. Cells synthesize proteins through a systematic procedure that begins in the nucleus when the gene code for a certain protein is *transcribed* from the cell's DNA into *messenger RNA,* or *mRNA.* The mRNA moves through nuclear pores to the rough endoplasmic reticulum (RER), where ribosomes *translate* the message one *codon* of three nucleotides, or *base pairs,* at a time. The ribosome uses *transfer RNA,* or *tRNA,* to fetch each required amino acid and then link them together through peptide bonds, also known as amide bonds, to form proteins (see Figure 2-4 for details).

TIP

Don't let the labels confuse you. Proteins are chains of amino acids (usually very long chains of at least 100 acids). Enzymes, used to catalyze reactions, also are chains of amino acids and therefore also are categorized as proteins. *Polypeptides,* or simply *peptides,* are shorter chains of amino acids used to bond larger protein molecules, but they also can be regarded as proteins. Both antibodies and hormones also are proteins, along with almost everything else in the body — hair, muscle, cartilage, and so on. Even the four basic blood types — A, B, AB, and O — are differentiated by the proteins (antigens) found in each.

38.–44. Fill in the blanks to complete the following sentences:

Protein synthesis begins in the cell's **38.** _____ when the gene for a certain protein is **39.** _____ into messenger RNA, or mRNA, which then moves on to the **40.** _____. There, ribosomes **41.** _____ three base pairs at a time, forming a series also referred to as a **42.** _____. Molecules called **43.** _____ then collect each amino acid needed so that the ribosomes can link them together through **44.** _____ bonds.

45. The word "protein" can refer to

a. hormones.

b. enzymes.

c. antibodies.

d. all of the above.

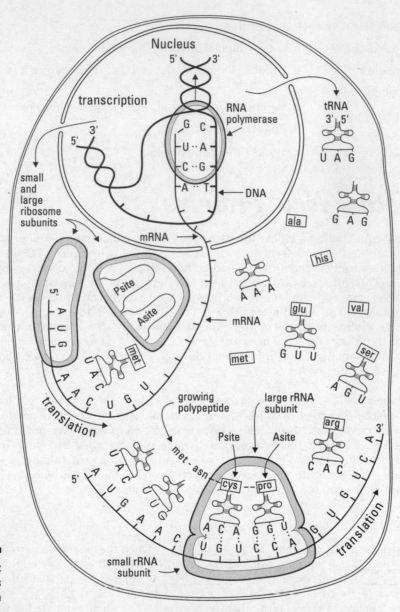

Figure 2-4:
The process
of protein
synthesis.

Protein Synthesis

© John Wiley & Sons, Inc.

46. Which of the following comes first in the protein-synthesis process?

a. Transfer RNA

b. Transcription

c. Peptide bonds

d. Translation

47. What is the difference between mRNA and tRNA?

a. mRNA is double-stranded.

b. tRNA contains deoxyribose.

c. mRNA originates in the cell's nucleus.

d. mRNA interacts with lysosomes.

48. What is a codon?

a. A sequence of three adjacent nucleotides on a DNA molecule

b. A type of amino acid used in the cellular production of protein

c. A specialized ribosome required to translate DNA

d. A base pair that connects complementary strands of DNA

Cycling Along: Grow, Rest, Divide, Die

The *cell life cycle,* usually referred to simply as the *cell cycle* or the *CDC (cell division cycle),* extends from the beginning of one cell division to the beginning of the next division. The human body produces new cells every day to replace those that are damaged or worn out.

The cell cycle is divided into two distinct phases:

✔ **Interphase:** Sometimes also called the resting stage, that label is a misnomer because the cell is actively growing and carrying out its normal metabolic functions as well as preparing for cell division.

✔ **Mitosis:** The period of cell division that produces new cells. (We cover this phase in detail in Chapter 3.)

New cells are produced for growth and to replace the billions of cells that stop functioning or are damaged beyond repair in the adult human body every day. Some cells, like blood and skin cells, are continually dividing because they have very short life cycles, sometimes only hours. Other cells, such as specialized muscle cells and certain nerve cells, may never divide at all.

49. How long can human cells live?

a. A few hours

b. A few days

c. Indefinitely

d. All of the above

50. The cell cycle is measured

a. by the number of times a cell divides.

b. from the beginning to the end of one cell division.

c. from the beginning of one cell division to the beginning of the next.

d. slowly, over time.

Answers to Questions on the Cell

The following are answers to the practice questions presented in this chapter.

1 – 3 The lipid bilayer structure of the cell membrane is made possible because phospholipid molecules contain two distinct regions: The **1. hydrophilic** region is attracted to water, and the **2. hydrophobic** region is repelled by water. Because it has both polar and nonpolar regions, a phospholipid is classified as a(n) **3. amphipathic** molecule.

4 What is osmosis? **c. The movement of water molecules through a semipermeable membrane.** Why not diffusion? Because diffusion has to do with the passive transport of substances *other than* water.

5 What is a hypotonic solution? **b. A solution that has a greater concentration of water than exists in the cell.**

The prefix *hypo* refers to under or below normal. The prefix *hyper* refers to excess, or above normal. Someone who has been out in the cold too long suffers hypothermia — literally insufficient heat. So a solution, or tonic, with very few particles would be hypotonic.

6 Injecting a large quantity of distilled water into a human's veins would cause many red blood cells to **a. swell and burst.** With a hypotonic solution outside the cell, the membrane would allow osmosis to continue past the breaking point.

7 The cell membrane does *not* function **c. in the production of energy.** Sometimes it may use energy in the form of ATP, but the cell membrane isn't involved directly in the production of energy.

8 What is a perinuclear cisterna? **a. A fluid-filled space within the nuclear envelope.** Note that all the other choices describe distinct parts of the cell: a ribosome, a mitochondrion, and a Golgi apparatus.

9 In the nucleus of a cell that isn't dividing, DNA **b. is packaged within chromatin threads.**

10 The nucleolus **d. assembles ribosomes.** It's not just a coincidence that the nucleolus sits at the heart of the genetic powerhouse.

11 What is the function of a mitochondrion? **b. To produce energy through aerobic respiration.** Each of the other options describes a function served by a different organelle.

12 Actin, which is the most abundant protein in human cells, **a. has filaments that contribute to the cytoskeleton.** It makes sense that the protein making up much of the cytoskeleton is the most abundant because the cytoskeleton accounts for up to 50 percent of the cell's volume.

13 Which organelle gets to take out the cellular trash? **c. Vacuole**

When it's time to clean house, you pull out the vacuum. Cells pull out the vacuoles.

14 Which organelle translates genetic information to direct protein synthesis? **a. The ribosome.**

When you think of a big protein-laden meal, you think of ribs. Ribs. Ribosome. Protein synthesis. Get it?

15 Which organelle has ribosomes attached to it? **c. Rough (granular) endoplasmic reticulum.**

16 Which organelle contains secretory materials? **a. Golgi apparatus.** The correct answer can't be the lysosome because that's already a vesicle, and it can't be the ribosome because you already know that synthesizes proteins.

17 Which of the following can change the consistency of cytoplasm? **d. All of the above.** In addition, molecular concentration and agitation also can change cytoplasmic consistency.

18 Organelles don't just float freely within a cell, some are found inside the nucleus. Which of these can be found within the nucleus? **c. Chromatin network.** Don't confuse the organelles in the cytoplasm with the organelles in the nucleus.

19-32 Following is how Figure 2-3, the cutaway view of the cell and its organelles, should be labeled.

19. **c. Cytoplasm**; 20. **n. Vesicle formation**; 21. **g. Nucleolus**; 22. **h. Nucleus**; 23. **m. Vacuole**; 24. **j. Ribosomes**; 25. **d. Golgi apparatus**; 26. **i. Plasma (cell) membrane**; 27. **b. Cilia**; 28. **f. Mitochondrion**; 29. **k. Rough endoplasmic reticulum**, 30. **l. Smooth endoplasmic reticulum**; 31. **e. Lysosome**; 32. **a. Centriole**

33 Mitochondrion: **d. Powerhouse of the cell**

34 Nucleolus: **e. Stores RNA in the nucleus**

35 Flagellum: **a. Long, whiplike organelle for locomotion**

36 Cytoplasm: **b. Fluidlike interior of the cell that may become a semisolid, or colloid**

37 Pysosomes: **c. Membranous sacs containing digestive enzymes**

38-44 Protein synthesis begins in the cell's **38. nucleus** when the gene for a certain protein is **39. transcribed** into messenger RNA, or mRNA, which then moves on to the **40. rough endoplasmic reticulum.** There, ribosomes **41. translate** three base pairs at a time, forming a series also referred to as a **42. codon.** Molecules called **43. tRNA** then collect each amino acid needed so that the ribosomes can link them together through **44. peptide bonds.**

45 The word "protein" can refer to **d. all of the above.** Proteins come in myriad shapes, sizes, and functions.

46 Which of the following comes first in the protein synthesis process? **b. Transcription.** Remember that you have to transcribe before you can translate.

47 What is the difference between mRNA and tRNA? **c. mRNA originates in the cell's nucleus.** Messenger RNA then moves to the rough endoplasmic reticulum where tRNA — transfer RNA — works with ribosomes.

48 What is a codon? **a. A sequence of three adjacent nucleotides on a DNA molecule.** It takes three nucleotides to round up a single amino acid.

49 Human cells can live **d. all of the above.** Cell life cycles can vary widely.

50 The cell cycle is measured **c. from the beginning of one cell division to the beginning of the next.**

Chapter 3

Divide and Conquer: Cellular Mitosis

● ●

In This Chapter
▶ Following the steps of cell division
▶ Understanding the results of errors in mitosis

● ●

Ever had so many places to be that you wished you could just divide yourself in two? Your cells already do that. Cell division is how one "mother" cell becomes two identical twin "daughter" cells. Cell division takes place for several reasons:

✔ **Growth:** Multicellular organisms, humans included, each start out as a single cell — the fertilized egg. That one cell divides (and divides and divides), eventually becoming an entire complex being.

✔ **Injury repair:** Uninjured cells in the areas surrounding damaged tissue divide to replace those that have been destroyed.

✔ **Replacement:** Cells eventually wear out and cease to function. Their younger, more functional neighbors divide to take up the slack.

✔ **Asexual reproduction:** No, human cells don't do this. Only single-celled organisms do.

Cell division occurs over the course of two processes: *mitosis,* which is when the chromosomes within the cell's nucleus duplicate to form two daughter nuclei; and *cytokinesis,* which takes place when the cell's cytoplasm divides to surround the two newly formed nuclei. Although cell division breaks down into several stages, there are no pauses from one step to another. Cell division as a whole is called mitosis because most of the changes occur during that process. Cytokinesis doesn't start until later. But mitosis and cytokinesis do end together.

Keep in mind: Cells are living things, so they mature, reproduce, and die. In this chapter, we review the cell cycle (as mitosis also is known), and you get plenty of practice figuring out what happens when and why.

Walking through the Mitotic Process

It may look like cells are living out their useful lives simply doing whatever specialized jobs they do best, but in truth mitosis is a continuous process. When the cell isn't actively splitting itself in two, it's actively preparing to do so. DNA and centrioles (arrays of microtubules; see Chapter 2) are being replicated, and the cell is bulking up on cytoplasm to make sure there's enough for both daughter cells. Mitosis may look like a waiting game, but there's plenty going on behind the scenes.

Waiting for action: Interphase

Interphase is the period when the cell isn't dividing. It begins when the new cells are done forming and ends when the cell prepares to divide. Although it's also called a "resting stage," there's constant activity in the cell during interphase.

Interphase is divided into subphases, each of which lasts anywhere from a few hours for those cells that divide frequently to days or years for those cells that divide less frequently (nerve cells, for example, can spend decades in interphase). The subphases are as follows:

- ✔ **G_1,** which stands for "gap" or "growth." During G_1, the cell creates its organelles, begins metabolism, grows, and synthesizes proteins.

- ✔ **S,** which stands for "synthesis." DNA synthesis or *replication* occurs during this subphase. The single double-helix DNA molecule inside the cell's nucleus becomes two new "sister" *chromatids,* and the centrosome (a type of organelle; see Chapter 2) is duplicated.

- ✔ **G_2,** which stands for "gap." Enzymes and proteins needed for cell division are produced during this subphase.

Sorting out the parts: Prophase

As the first active phase of mitosis, *prophase* is when structures in the cell's nucleus begin to disappear, including the nuclear membrane (or envelope), nucleoplasm, and nucleoli. The two centrosomes, duplicated in the synthesis process during interphase and each containing two centrioles, push apart to opposite ends of the nucleus, forming poles. The centrioles produce protein filaments that form *mitotic spindles* between the poles as well as *asters* (or *astral rays*) that radiate from the poles into the cytoplasm.

At the same time, the *chromatin* threads (or *chromonemata*) shorten and coil, forming visible chromosomes. The chromosomes divide into *chromatids* that remain attached at an area called the *centromere,* which produces microtubules called *kinetochore* fibers. These interact with the mitotic spindles to assure that each daughter cell ultimately has a full set of chromosomes. The *chromatids* start to migrate toward the *equatorial plane,* an imaginary line between the poles.

Dividing at the equator: Metaphase

After the chromosomes are lined up and attached along the cell's newly formed equator, *metaphase* officially debuts. The nucleus itself is gone. The chromatids line up exactly along the centerline of the cell (or the equatorial plane), attaching to the mitotic spindles by the centromeres. The centromere also is attached by microtubules (spindles) to opposite poles in the cell.

Packing up to move out: Anaphase

In *anaphase,* the centromeres split, separating the duplicate chromatids and forming two chromosomes. The spindles attached to the divided centromeres shorten, pulling the chromosomes toward the opposite poles. The cell begins to elongate. In late anaphase, as the chromosomes approach the poles, a slight furrow develops in the cytoplasm, showing where cytokinesis will eventually take place.

Pinching off: Telophase

Telophase occurs as the chromosomes reach the poles and the cell nears the end of division. The spindles and asters of early mitosis disappear, and each newly forming cell begins to synthesize its own structure. New nuclear membranes enclose the separated chromosomes. The coiled chromosomes unwind, becoming chromonemata once again. There's a more pronounced pinching, or furrowing, of the cytoplasm into two separate bodies, but there continues to be only one cell.

Splitting up: Cytokinesis

Cytokinesis means it's time for the big breakup. The furrow, formed by a contractile ring that will divide the newly formed sister nuclei, migrates inward until it cleaves the single, altered cell into two new cells. Each new cell is smaller and contains less cytoplasm than the mother cell, but the daughter cells are genetically identical to each other and to the original mother cell, and will grow to normal size during interphase.

Try this warm-up question:

Q. True or false: Nothing happens during interphase.

A. False. While it's also called a "resting stage" (a misnomer!), plenty is going on during interphase.

1. Cells are dormant during interphase.

   a. True

   b. False

2. The G_1 subphase of interphase is

   a. the period of DNA synthesis.

   b. the most active phase.

   c. the phase between S and G_2.

   d. part of cell division.

3. What happens to DNA during interphase?

   a. It's duplicated during the S subphase.

   b. It's cut in two during the G_2 subphase.

   c. It unwinds rapidly during the B subphase.

   d. It develops a protective film during the G_1 subphase.

4. During prophase, the nuclear membrane, or envelope,

    a. splits into thirds.

    b. develops a series of concave dimples.

    c. begins to disappear.

    d. changes color.

5. Which of the following happens in prophase?

    a. The chromatids align on the equatorial plane.

    b. The chromosomes divide into chromatids.

    c. The nucleus reappears.

    d. The chromosomes move to opposite poles.

6. Which of the following is true for metaphase?

    a. The nuclear membrane appears.

    b. The chromosomes move to the poles.

    c. The chromatids align on the equatorial plane.

    d. It's composed of subphases G_1, S, and G_2.

7. During metaphase, each chromosome consists of two duplicate chromatids.

    a. True

    b. False

8. Identify an event that does *not* happen during anaphase.

    a. Early cytokinesis occurs with slight furrowing.

    b. The cell goes through subphase G_1.

    c. Spindles shorten.

    d. The centromeres split.

9. During anaphase, genetically identical chromosomes

    a. move toward each other.

    b. link with mitochondria.

    c. are pulled toward opposite poles.

    d. multiply rapidly.

10. Which event does *not* occur during telophase?

    a. The chromosomes uncoil.

    b. The chromosomes reach the poles.

    c. The chromosomes become more distinct.

    d. The nuclear membrane reforms.

11. What structures disappear during telophase?

   a. Spindles and asters

   b. Nuclear membranes

   c. Nucleolei

   d. Chromonemata

12. Which is the correct order of phases of the cell cycle and mitosis?

   a. Prophase, metaphase, telophase, anaphase

   b. Prophase, metaphase, anaphase, telophase

   c. Metaphase, anaphase, telophase, prophase

   d. Anaphase, metaphase, telophase, prophase

13.–24. Use the terms that follow to identify the stages and cell structures shown in Figure 3-1.

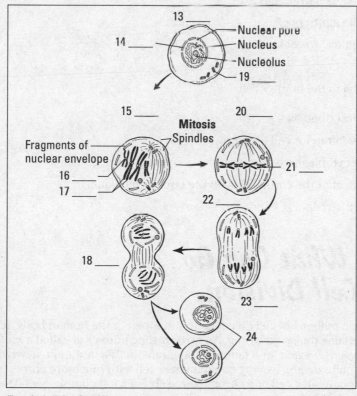

Figure 3-1: Cell structures and changes that make up the stages of mitosis.

Illustration by Kathryn Born, MA

    a. Anaphase

    b. Centromere

    c. Daughter cells

    d. Chromatin

    e. Cytokinesis

    f. Telophase

    g. Interphase

    h. Chromosomes aligned at equator

    i. Metaphase

    j. Centriole

    k. Prophase

    l. Chromatids

25. The two newly formed daughter cells are

    a. the same size as the mother cell.

    b. not genetically identical to each other.

    c. unequal in size.

    d. genetically identical to the mother cell.

26. Cytokinesis can be described as

    a. the period of preparation for cell division.

    b. the dividing of the cytoplasm to surround the two newly formed nuclei.

    c. the stage of alignment of the chromatids on the equatorial plane.

    d. the initiation of cell division.

Understanding What Can Go Wrong during Cell Division

With the millions upon millions of cell divisions that happen in the human body, it's not surprising that sometimes things go wrong. An error during mitosis is called a *mutation*. One kind of mutation is *nondisjunction,* or a failure to separate. In this mutation, newly formed chromosomes don't quite divide, leaving one daughter cell with one more chromosome than normal and the other daughter cell one chromosome shy of a full complement. Down's syndrome is an example of what happens when nondisjunction occurs. A normal human cell has 46 chromosomes, but that of a Down's sufferer has 47.

Mitosis can also end up on fast-forward. Accelerated mitosis can lead to the formation of a tumor, also called a *neoplasm.* The rate of division usually restricts itself to replacing worn out or injured cells, but with accelerated mitosis, the cells don't know when to stop dividing.

27. What is the name of the mutation during mitosis that can result in Down's syndrome?

    a. Neoplasm

    b. Nondisjunction

    c. Nondivision

    d. Acceleration

28. Accelerated mitosis can cause a neoplasm, which is also known as a

    a. nondisjunction.

    b. daughter cell.

    c. tumor.

    d. complement.

Answers to Questions on Mitosis

The following are answers to the practice questions presented in this chapter.

1 Cells are dormant during interphase. **b. False.** Cells are at their most active during interphase.

2 The G_1 subphase of interphase is **b. the most active phase.** With all that organelle-growing, metabolizing, and protein-synthesizing that takes place during G_1, this isn't surprising.

3 What happens to DNA during interphase? **a. It's duplicated during the S subphase.** Remember, the cell is S-ynthesizing new DNA molecules during this phase.

4 During prophase, the nuclear membrane, or envelope, **c. begins to disappear.**

5 Which of the following happens in prophase? **b. The chromosomes divide into chromatids.** But don't forget that they remain attached at the centromere.

6 Which of the following is true for metaphase? **c. The chromatids align on the equatorial plane.** Each of the other answer choices occurs during earlier or later phases.

7 During metaphase, each chromosome consists of two duplicate chromatids. **a. True.** That way, each resulting daughter cell will have identical chromosomes.

8 Identify an event that does *not* happen during anaphase. **b. The cell goes through subphase G_1.** G_1 took place back in interphase.

9 During anaphase, genetically identical chromosomes **c. are pulled toward opposite poles.** As the cell nears the end of division, it makes sense that duplicate packages move to opposite ends of the cell.

10 Which event does *not* occur during telophase? **c. The chromosomes become more distinct.** That change happened back in prophase.

11 What structures disappear during telophase? **a. Spindles and asters.** These structures disappear because they're no longer needed at the end of mitosis.

12 Which is the correct order of phases of the cell cycle and mitosis? **b. Prophase, metaphase, anaphase, telophase.**

13-24 Following is how Figure 3-1, the stages and structures of mitosis, should be labeled.

13. **g. Interphase;** 14. **d. Chromatin;** 15. **k. Prophase;** 16. **b. Centromere;** 17. **l. Chromatids;** 18. **f. Telophase;** 19. **j. Centriole;** 20. **i. Metaphase;** 21. **h. Chromosomes aligned at equator;** 22. **a. Anaphase;** 23. **e. Cytokinesis;** 24. **c. Daughter cells**

25 The two newly formed daughter cells are **d. genetically identical to the mother cell.** None of the other answer options makes sense.

26 Cytokinesis can be described as **b. the dividing of the cytoplasm to surround the two newly formed nuclei.**

27 What is the name of the mutation during mitosis that can result in Down's syndrome? **b. Nondisjunction.** Split the word into pieces. "Non" means negative, or it didn't happen. "Dis" means "apart" or "asunder." And you know that a "junction" is where things are joined. Down's syndrome occurs when chromosomes fail to split up properly, leaving daughter cells with the wrong number of chromosomes.

28 Accelerated mitosis can cause a neoplasm, which is also known as a **c. tumor.**

Chapter 4

The Study of Tissues: Histology

Oh, what tangled webs we weave! As the chapter title says, *histology* is the study of tissues, but you may be surprised to find out that the Greek *histo* doesn't translate as "tissue" but instead as "web." It's a logical next step after reviewing the cell and cellular division (see Chapters 2 and 3) to take a look at what happens when groups of similar cells "web" together to form tissues. The four different types of tissue in the body are as follows:

✔ Epithelial tissue (from the Greek *epi–* for "over" or "outer")

✔ Connective tissue

✔ Muscle tissue

✔ Nerve tissue

In this chapter, you find a quick review of the basics of each of these types of tissues along with practice questions to test your knowledge of them.

Getting into Your Skin: Epithelial Tissue

Perhaps because of its unique job of both protecting the outer body and lining internal organs, epithelial tissue has many characteristics that distinguish it from other tissue types.

Epithelial tissues, which generally are arranged in sheets or tubes of tightly packed cells, always have a free, or *apical,* surface that can be exposed to the air or to fluid. That free surface also can be covered by additional layers of epithelial tissue. But whether it's layered or not, each epithelial cell has *polarity* (a top and a bottom), and all but one side of the surface cells are tucked snugly against neighboring cells. The apical side sometimes has cytoplasmic projections such as *cilia,* hairlike growths that can move material over the cell's surface, or *microvilli,* fingerlike projections that increase the cell's surface area for increased adsorption and more efficient diffusion of absorbents and secretions. Opposite the apical side is the *basal* side (think basement), which typically attaches to some kind of connective tissue.

*Ad*sorption involves the *ad*hesion of something to a surface, while *ab*sorption involves a fluid permeating or dissolving into a liquid or solid.

Epithelial tissue serves several key functions, including the following:

- ✔ **Protection:** Skin protects vulnerable structures or tissues deeper in the body.

- ✔ **Barrier:** Epithelial tissues prevent foreign materials from entering the body and retain interstitial fluids.

- ✔ **Sensation:** Sensory nerve endings embedded in epithelial tissue connect the body with outside stimuli.

- ✔ **Secretion:** Epithelial tissue in glands can be specialized to secrete enzymes, hormones, and fluids.

- ✔ **Absorption:** Linings of the internal organs are specialized to facilitate absorption of materials into the body.

- ✔ **Filtration:** The epithelium of the respiratory tract protects the body by filtering out dirt and particles and cleaning the air that's inhaled. The blood is filtered in epithelial tissue in the kidneys.

- ✔ **Diffusion:** Simple squamous epithelial cells form a semipermeable membrane that allows selective diffusion of materials to pass through under osmotic pressure, which contributes to the filtration function. Diffusion is also involved in transport processes at the tissue interface in absorption and secretion functions.

Single-layer epithelial tissue is classified as *simple*. Tissue with more than one layer is called *stratified*. Epithelial tissues also can be classified according to shape:

- ✔ **Squamous:** *Squamous* cells are thin and flat.

- ✔ **Cuboidal:** As the name implies, *cuboidal* cells are equal in height and width and shaped like cubes.

- ✔ **Columnar:** *Columnar* cells are taller than they are wide.

Following are the ten primary types of epithelial tissues:

- ✔ **Simple squamous epithelium:** This flat layer of scalelike cells looks like a fish's scales and is useful in diffusion, secretion, or absorption. Each cell nucleus is centrally located and is round or oval. Simple squamous epithelium lines the lungs' air sacs where oxygen and carbon dioxide are exchanged.

- ✔ **Simple cuboidal epithelium:** These cube-shaped cells, found in a single layer that looks like a microscopic mattress, have centrally located nuclei that usually are round. They are found in the ovaries, line the ascending limb of the loop of Henle and the convoluted tubules in the kidney, and line some glands such as the thyroid, sweat glands, and salivary glands, as well as the inner surface of the eardrum, known as the *tympanic membrane*. Functions of this type of epithelium are secretion, absorption, and tube formation. The kidney tubules have microvilli that increase the area of adsorption for more efficient diffusion and absorption.

✔ **Simple columnar epithelium:** These densely packed cells are taller than they are wide, with nuclei located near the base of each cell. Found lining the digestive tract from the stomach to the anal canal, the functions of this type of epithelium are secretion and absorption.

✔ **Simple columnar ciliated epithelium:** A close cousin to simple columnar epithelium, this type of tissue has hairlike cilia that can move mucus and other substances across the cell. It's found lining the small respiratory tubes.

✔ **Pseudostratified columnar epithelium:** Pay attention to the prefix *pseudo–* here, which means "false." It may look multilayered because the cells' nuclei are scattered at different levels, but it's not. This type of epithelium is found in the large ducts of the parotid glands (salivary glands) and some segments of the male reproductive system, including the urethra.

✔ **Pseudostratified ciliated columnar epithelium:** Another variation on a theme, this tissue is nearly identical to pseudostratified columnar epithelium. The difference is that this tissue's free surface has cilia, making it ideal for lining air passages because the cilia's uniform waving action causes dust and dirt particles trapped in a thin layer of mucus produced by interspersed mucus cells to move in one direction — away from the lungs and toward the throat and mouth. The surface of the mucus-producing cells scattered among the ciliated epithelial cells is covered with microvilli. The microvilli occur on epithelial surfaces where absorption and secretion take place.

✔ **Stratified squamous epithelium:** This tissue is the stuff you see everyday — your outer skin, or epidermis. This multilayered tissue has squamous cells on the outside plus deeper layers of cuboidal or columnar cells. Found in areas where the outer cell layer is constantly worn away, this type of epithelium regenerates its surface layer with cells from lower layers.

✔ **Stratified cuboidal epithelium:** This multilayered epithelium can be found in sweat glands, conjunctiva of the eye, and the male urethra. Its function is primarily protection.

✔ **Stratified columnar epithelium:** Also multilayered, this epithelium is found lining parts of the male urethra, excretory ducts of glands, and some small areas of the anal mucous membrane.

✔ **Stratified transitional epithelium:** This epithelium is referred to as *transitional* because its cells can shape-shift from cubes to squamouslike flat surfaces and back again. Found lining the urinary bladder, the cells stretch and flatten out to make room for urine. When urination occurs, the cells relax and assume their original form.

Following are some practice questions dealing with epithelial tissue.

Q. Stratified epithelial tissue can be described as

a. a thin sheet of cells.

b. covered in cilia.

c. layers of stacked epithelial cells.

d. a long string of tissue.

A. The correct answer is layers of stacked epithelial cells. Remember that "stratified" means layers.

1. How are epithelial cells shaped?

   a. Like columns

   b. Like cubes

   c. Thin and flat

   d. All of the above

2. How is epithelial tissue classified?

   a. By number of layers

   b. By composition of matrix

   c. By cell shape

   d. By both the number of layers and the cell shape

3. Stratified transitional epithelial tissue

   a. primarily protects specific areas.

   b. has the ability to stretch.

   c. is constantly worn away.

   d. is found in the salivary glands.

4.–8. Match the epithelial tissue with its location in the body.

   4. _____ Simple columnar   **a.** Urinary bladder

   5. _____ Stratified squamous   **b.** Tubules of the kidney

   6. _____ Stratified transitional   **c.** Digestive tract

   7. _____ Pseudostratified ciliated columnar   **d.** Epidermis of the skin

   8. _____ Simple cuboidal   **e.** Respiratory passages

9. The key functions of epithelial tissue do *not* include

   a. protection.

   b. secretion.

   c. contraction.

   d. sensation.

10.–19. Use the terms that follow to identify the epithelial tissues shown in Figure 4-1.

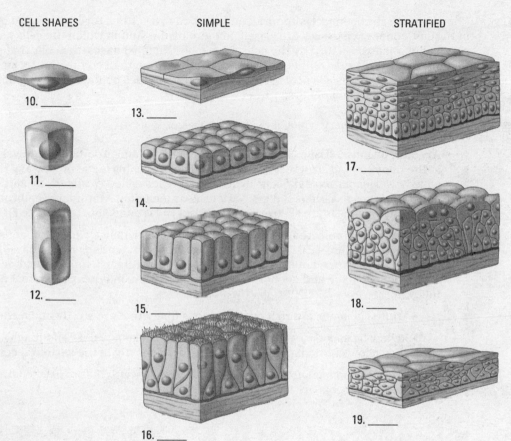

CELL SHAPES   SIMPLE   STRATIFIED

10. _____

11. _____

12. _____

13. _____

14. _____

15. _____

16. _____

17. _____

18. _____

19. _____

Figure 4-1:
Epithelial
tissues.

Illustration by Imagineering Media Services Inc.

a. Stratified squamous

b. Simple columnar

c. Squamous

d. Stratified transitional stretched

e. Simple squamous

f. Columnar

g. Pseudostratified ciliated columnar

h. Cuboidal

i. Stratified transitional relaxed

j. Simple cuboidal

Making a Connection: Connective Tissue

Connective tissues connect, support, and bind body structures together. Unlike other types of tissues, connective tissues are classified more by the stuff in which the cells lie — the extracellular matrix — than by the cells themselves. In most cases, the cells that produce that matrix are scattered within it. The load-bearing strength of connective tissue comes from a fibrous protein called *collagen.* All connective tissues contain a varying mix of collagen, elastic, and reticular fibers.

Following are the primary types of connective tissue:

✔ **Areolar, or loose, tissue:** This tissue exists between and around almost everything in the body to bind structures together and fill space. It's made up of wavy ribbons called *collagenous protein fibers,* cylindrical threads called *elastic fibers,* and *amorphous ground substance,* a semisolid gel. Various cells including lymphocytes, fibroblasts, fat cells, and mast cells are scattered throughout the ground substance (see Figure 4-2).

✔ **Dense regular connective tissue:** Made up of parallel, densely packed bands or sheets of fibers (see Figure 4-2), this type of tissue is found in tendons as bundles of collagenous fibers attaching muscles to bone and in ligaments extending from bone to bone, surrounding a joint, and anchoring organs. The two main types of ligament are the following:

• White ligaments are rich in collagenous fibers that are sturdy and inelastic.

• Yellow ligaments *(ligamenta flava)* are rich in elastic fibers and tough, although they allow elastic movement. They're found mostly in the vertebral column.

Dense regular connective tissue has great tensile strength that resists lengthwise pulling forces. Ligaments are more elastic than tendons.

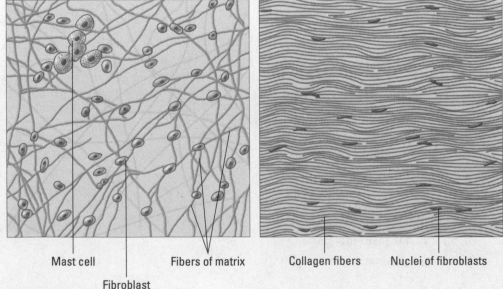

Figure 4-2:
Areolar
tissue and
dense
regular
connective
tissue.

Mast cell   Fibers of matrix   Collagen fibers   Nuclei of fibroblasts

Fibroblast

Illustration by Imagineering Media Services Inc.

- **Dense irregular connective tissue:** Also known as *dense fibrous connective tissue,* it consists of fibers that twist and weave around each other, forming a thick tissue that can withstand stresses applied from any direction. This tissue makes up the strong inner skin layer called the *dermis* as well as the outer capsule of organs like the kidney and the spleen.

- **Adipose tissue:** Composed of fat cells, this tissue forms padding around internal organs, reduces heat loss through the skin, and stores energy in fat molecules called triglycerides. Fat molecules fill the cells, forcing the nuclei against the cell membranes and giving them a ringlike shape. Adipose has an *intracellular matrix* rather than an extracellular matrix.

- **Reticular tissue:** Literally translated as "weblike" or "netlike," reticular tissue is made up of slender, branching reticular fibers with reticular cells overlaying them. Its intricate structure makes it a particularly good filter, which explains why it's found inside the spleen, lymph nodes, and bone marrow.

- **Cartilage:** These firm but flexible tissues, made up of collagen and elastic fibers, have no nerve cells or blood vessels (a state called *nonvascular* or *avascular*). Cartilage contains openings called *lacunae* (from the Latin word *lacus* for "lake" or "pit") that enclose mature cells called *chondrocytes,* which are preceded by cells called *chondroblasts.* A membrane known as the *perichondrium* surrounds cartilage tissue, which also contains a gelatinous protein called *chondrin.* There are three types of cartilage:

  - **Hyaline cartilage:** The most abundant cartilage in the body, it's collagenous and made up of a uniform matrix pocked with chondrocytes. It lays the foundation for the embryonic skeleton, forms the rib (or costal) cartilages, makes up nose cartilage, and covers the articulating surfaces of bones.

  - **Fibrocartilage:** As the name implies, fibrocartilage contains thick, compact collagen fibers. The spongelike structure, with the lacunae and chondrocytes lined up within the fibers, makes it a good shock absorber. It's found in the intervertebral discs of the vertebral column and in the symphysis pubis at the front of the pelvis.

  - **Elastic cartilage:** Similar to hyaline cartilage but with a much greater abundance of elastic fibers, elastic cartilage has more tightly packed lacunae and chondrocytes between parallel elastic fibers. This cartilage — which makes up structures where a specific form is important such as the outer ear, Eustachian tube, and epiglottis — tends to bounce back to its original shape after being bent.

- **Bone, or osseous, tissue:** Essentially, bone is mineralized connective tissue formed into repeating patterns called *Haversian systems* or *osteons.* In the center of each system is a large opening, the *Haversian canal,* that contains blood vessels, lymph vessels, and nerves. The central canal is surrounded by thin layers of bone called *lamellae* that contain the lacunae, which in turn contain *osteocytes* (bone cells). Smaller *canaliculi* connect the lacunae and circulate tissue fluids from the blood vessels to nourish the osteocytes. (We explore bone in more detail in Chapter 5.)

- **Blood:** Yes, blood is considered a type of connective tissue. Like other connective tissues, it has an extracellular matrix — in this case, plasma — in which *erythrocytes* (red blood cells), *leukocytes* (white blood cells), and *thrombocytes* (platelets) are suspended. (Blood also is considered a vascular tissue because it circulates inside arteries and veins, but we get into that in Chapter 10.) Roughly half of blood's volume is fluid or plasma while the other half is suspended cells. Erythrocytes are concave on both sides and contain a pigment, *hemoglobin,* which supplies oxygen to the body's cells and takes carbon dioxide away. There are approximately 5 million erythrocytes per cubic millimeter of whole blood. Thrombocytes, which number approximately 250,000 per cubic millimeter, are fragments of cells used in blood clotting. Some leukocytes are large *phagocytic* cells (literally "cells that eat") that are part of the body's immune system. There are, however, relatively few of them — less than 10,000 per cubic millimeter.

20. What is adipose tissue composed of?

   a. Mast cells

   b. Chondrocytes

   c. Osteocytes

   d. Fat cells

21. What are tendons composed of?

   a. Elastic tissue

   b. Dense regular connective tissue

   c. Areolar tissue

   d. Fibrocartilage

22. Hyaline cartilage

   a. covers the surface of articulating bones.

   b. lines the symphysis pubis.

   c. connects to the epiglottis.

   d. cushions intervertebral discs.

23. Why does dense irregular connective tissue twist and weave around?

   a. To fill gaps between other tissues

   b. To withstand stresses applied from any direction

   c. To prevent dead or dying cells from weakening its structure

   d. To provide firm connections between muscles and bones

24. Which tissue contains lacunae with osteocytes?

   a. Elastic cartilage

   b. Bone

   c. Hyaline cartilage

   d. Blood

25. What function do thrombocytes serve?

   a. They block invading microbes.

   b. They carry oxygen to connective tissues.

   c. They nourish osteocytes.

   d. They contribute to blood clotting.

Flexing It: Muscle Tissue

Although we review how muscles work in Chapter 6, in histology you should know that muscle tissue is made up of fibers known as *myocytes*. The cytoplasm within the fibers is called *sarcoplasm,* and within that sarcoplasm are minute *myofibrils* that contain the protein filaments *actin* and *myosin*. These filaments slide past each other during a muscle contraction, shortening the fiber.

Following are the three types of muscle tissue (see Figure 4-3):

✔ **Smooth muscle tissue:** This type of tissue contracts without conscious control. Made up of spindle-shaped fibers with large, centrally located nuclei, it's found in the walls of internal organs, or *viscera*. Smooth muscle gets its name from the fact that, unlike other muscle tissue types, it is not striated.

✔ **Cardiac muscle tissue:** This tissue is composed of cylindrical branching fibers, *myocardiocytes,* each with a central nucleus and alternating light and dark striations. Between the fibers are dark structures called *intercalated discs* that hold the fibers together. As with smooth muscle, cardiac muscle tissue contractions occur through the autonomic nervous system (involuntary control).

✔ **Skeletal, or striated, muscle tissue:** Biceps, triceps, pecs — these are the muscles that bodybuilders focus on. As the name implies, skeletal muscles attach to the skeleton and are controlled throughout by the conscious (voluntary) control function of the central nervous system for movement. Muscle fibers are cylindrical with several nuclei in each cell (which makes them *multinucleated*) and cross-striations throughout.

Figure 4-3:
Muscle
tissues:
Smooth,
cardiac, and
skeletal.

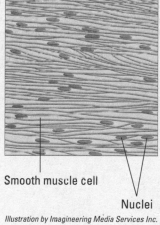  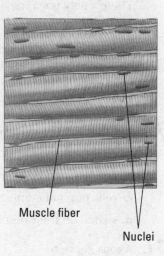

Smooth muscle cell   Intercalated disc   Muscle fiber

Nuclei   Nucleus   Nuclei

Illustration by Imagineering Media Services Inc.

26. Which type of tissue is multinucleated?

a. Skeletal muscle tissue

b. Cardiac muscle tissue

c. Smooth muscle tissue

d. Autonomic muscle tissue

27. Intercalated discs can be found in which type of muscle?

a. Cardiac

b. Skeletal

c. Smooth

d. Striated

28. Skeletal muscle tissue has prominent lines across the fiber. What are they called?

   a. Fibroblasts

   b. Multinucleation

   c. Lacunae

   d. Striations

29. Which of the following contains smooth muscle tissue?

   a. The heart

   b. The urinary bladder

   c. The bicep

   d. The deltoid

Getting the Signal Across: Nerve Tissue

There's only one type of nerve tissue and only one primary type of cell in it: the *neuron*. A neuron is unique in that it can both generate and conduct electrical signals in the body. That process starts when sense receptors receive a stimulus that causes electrical impulses to be sent through rootlike cytoplasmic projections called *dendrites*. From there, the impulse moves through the body of the cell and into another type of cytoplasmic projection called an *axon* that hands the signal off chemically to the next cell down the line. (We look more closely at how all that happens when we examine the central nervous system in Chapter 15.)

30. Neurons are the only cells in the body that can

   a. change shape in response to stimuli.

   b. produce and transmit electrical impulses.

   c. seek out and connect with other types of cells.

   d. switch off their replication process.

31. Axons serve to

   a. generate impulses in response to stimuli.

   b. receive impulses into the cell body.

   c. transmit impulses away from the cell body.

   d. project dendrites into the spine.

32. Dendrites serve to

   a. receive impulses.

   b. transmit impulses.

   c. generate impulses.

   d. amplify impulses.

Answers to Questions on Histology

The following are answers to the practice questions presented in this chapter.

1 Epithelial cells can be shaped like **d. all of the above.** Whether it's cuboidal, columnar, or flat and thin, it's still an epithelial cell.

2 Epithelial tissue is classified by **d. both the number of layers and the cell shape.** Classification requires looking at both simultaneously.

3 Stratified transitional epithelial tissue **b. has the ability to stretch.** This tissue lines the bladder, so it had better be stretchy!

4 Simple columnar: **c. Digestive tract**

5 Stratified squamous: **d. Epidermis of the skin**

6 Stratified transitional: **a. Urinary bladder**

7 Pseudostratified ciliated columnar: **e. Respiratory passages**

8 Simple cuboidal: **b. Tubules of the kidney**

9 The key functions of epithelial tissue do *not* include **c. contraction.** Muscles contract; epithelial tissue does not.

10–19 Following is how Figure 4-1, the types of epithelial cells and tissues, should be labeled.

10. **c. Squamous**; 11. **h. Cuboidal**; 12. **f. Columnar**; 13. **e. Simple squamous**; 14. **j. Simple cuboidal**; 15. **b. Simple columnar**; 16. **g. Pseudostratified ciliated columnar**; 17. **a. Stratified squamous**; 18. **i. Stratified transitional relaxed**; 19. **d. Stratified transitional stretched**

20 Adipose tissue is composed of **d. fat cells.** Think of the Latin *adeps,* which means "fat."

21 Tendons are composed of **b. dense regular connective tissue.** Tendons are dense and exhibit a regular pattern.

22 Hyaline cartilage **a. covers the surface of articulating bones.** It's all in the names of the three types of cartilage. Hyaline comes from the Greek word for *glass,* which seems appropriate for something covering a surface. Fibrocartilage contains thick bunches of fibers, which doesn't sound right for covering a surface. And elastic cartilage tends to be found where exact shapes are important, such as in the outer ear.

23 Why does dense irregular connective tissue twist and weave around? **b. To withstand stresses applied from any direction.**

24 Which tissue contains lacunae with osteocytes? **b. Bone.** Remember that *osteo* is the Latin word for "bone."

25 What function do thrombocytes serve? **d. They contribute to blood clotting.** Knowing that the Greek word *thrombos* means "clot" can help you spot the correct answer in this question.

26 Which type of tissue is multinucleated? **a. Skeletal muscle tissue.** Cardiac muscle tissue and smooth muscle tissue typically have only one nucleus per cell; autonomic muscle tissue is a trick answer choice.

27 Intercalated discs can be found in which type of muscle? **a. Cardiac.** Intercalated discs, as you should or will know from studying the circulatory system, are involved in conducting signals for the heart to pump.

28 Skeletal muscle tissue has prominent lines across the fiber. What are they called? **d. Striations.** Striations — think *stri*pes — are light and dark lines.

29 Which of the following contains smooth muscle tissue? **b. The urinary bladder.** The other answer choices contain striated tissue, which technically means that they aren't smooth.

30 Neurons are the only cells in the body that can **b. produce and transmit electrical impulses.** Sounds rather nervy of them, don't you think?

31 Axons serve to **c. transmit impulses away from the cell body.** Each neuron cell usually has only one axon, although it may branch off several times. Axons transmit a signal away, while dendrites receive a signal.

32 Dendrites serve to **a. receive impulses.** Usually, there are several dendrites per neuron cell.

Part II

Weaving It Together: Bones, Muscles, and Skin

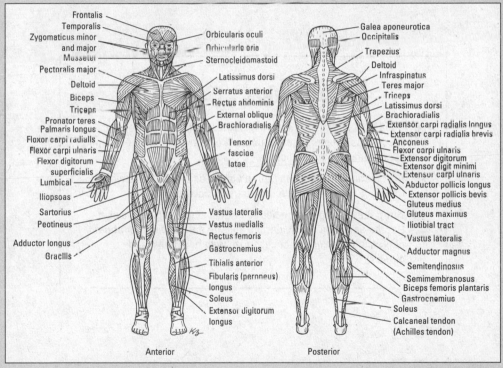

Anterior · Posterior

Illustration by Kathryn Born, MA

In this part . . .

✔ Build on a strong foundation: Explore how bones are formed. Piece those bones together into a skeleton and distinguish your axial from your appendicular.

✔ Articulate some joints and attach muscles to that framework. Contract and relax to leverage muscle power.

✔ Wrap the package in the body's largest single organ: the skin. Get the scoop on the epidermis and the dermis; find out about nerves in the skin; and accessorize with hair, nails, and glands.

Chapter 5

A Scaffold to Build On: The Skeleton

In This Chapter

▶ Getting to know what your bones do

▶ Understanding classifications, structures, and ossification

▶ Keeping the axial skeleton in line

▶ Checking out the appendicular skeleton

▶ Playing with joints

Human *osteology*, from the Greek word for "bone" (*osteon*) and the suffix –*logy*, which means "to study," focuses on the 206 bones in the adult body endoskeleton. But osteology is more than just bones; it's also ligaments and cartilage and the joints that make the whole assembly useful. In this chapter, you get lots of practice exploring the skeletal functions and how the joints work together.

Understanding the Functions of Dem Bones

The skeletal system as a whole serves five key functions:

▸ **Protection:** The skeleton encases and shields delicate internal organs that might otherwise be damaged during motion or crushed by the weight of the body itself. For example, the skull's cranium houses the brain, and the ribs and sternum of the thoracic cage protect organs in the thoracic body cavity.

▸ **Movement:** By providing anchor sites and a scaffold against which muscles can contract, the skeleton makes motion possible. The bones act as levers, the joints are the fulcrums, and the muscles apply the force. For instance, when the biceps muscle contracts, the radius and ulna bones of the forearm are lifted toward the humerus bone of the upper arm.

▸ **Support:** The vertebral column's curvatures play a key role in supporting the entire body's weight, as do the arches formed by the bones of the feet. Upper body support flows from the *clavicle*, or collarbone, which is the only bone that attaches the upper extremities to the axial skeleton and the only horizontal long bone in the human body.

▸ **Mineral storage:** Calcium, phosphorous, and other minerals like magnesium must be maintained in the bloodstream at a constant level, so they're "banked" in the bones in case the dietary intake of those minerals drops. The bones' mineral content is constantly renewed, refreshing entirely about every nine months. A 35 percent decrease in blood calcium will cause convulsions.

> ✔ **Blood cell formation:** Called *hemopoiesis* or *hematopoiesis,* most blood cell formation takes place within the red marrow inside the ends of long bones as well as within the vertebrae, ribs, sternum, and cranial bones. Marrow produces three types of blood cells: *erythrocytes* (red cells), *leukocytes* (white cells), and *thrombocytes* (platelets). Most of these are formed in red bone marrow, although some types of white blood cells are produced in fat-rich yellow bone marrow. At birth, all bone marrow is red. With age, it converts to the yellow type. In cases of severe blood loss, the body can convert yellow marrow back to red marrow in order to increase blood cell production.

The following is an example of a question dealing with skeletal functions:

Q. Which of the following is not a function of the skeleton?

   a. Support of soft tissue

   b. Hemostasis

   c. Production of red blood cells

   d. Movement

A. The correct answer, of course, is hemostasis, which is the stoppage of bleeding or blood flow. This is one of those frequent times when study of anatomy and physiology boils down to rote memorization of Latin and Greek roots (for help, check out the Cheat Sheet at www.dummies.com/cheatsheet/anatomyphysiologywb).

1. What is happening during hematopoiesis?

   a. Bone marrow is converting from red type to yellow type.

   b. Bone marrow is forming red blood cells.

   c. Bone marrow is converting back from yellow type to red type.

   d. Bone marrow is releasing minerals into the blood stream.

2. Why do the bones contain minerals like phosphorous, calcium, and magnesium?

   a. To enhance the skeleton's structural integrity

   b. To provide cushioning and nourishment to the joints

   c. To ensure availability of these minerals, even if insufficient amounts of them are consumed

   d. To form networks within the bones' structure

3. In what way does the skeleton make locomotion possible?

   a. It ensures hemopoiesis occurs as needed.

   b. Muscles use the skeleton as a scaffold against which they can contract.

   c. It cushions the joints against erythrocytes.

   d. It ensures the thoracic cavity remains properly filled.

4. The curvatures in some bone structures serve the following purpose:

   a. Support the body

   b. Make the body more flexible

   c. Enhance circulation throughout the body

   d. Divide different body areas

Boning Up on Classifications, Structures, and Ossification

Adult bones are composed of 30 percent protein (called *ossein*), 45 percent minerals (including calcium, phosphorus, and magnesium), and 25 percent water. Minerals give the bone strength and hardness. At birth, bones are soft and pliable because of cartilage in their structure. As the body grows, older cartilage gradually is replaced by hard bone tissue. Minerals in the bones increase with age, causing the bones to become more brittle and easily fractured.

Various types of bone make up the human skeleton, but fortunately for memorization purposes, bone type names match what the bones look like. They are as follows:

- **Long bones,** like those found in the arms and legs, form the weight-bearing part of the skeleton.

- **Short bones,** such as those in the wrists (carpals) and ankles (tarsals), have a blocky structure and allow for a greater range of motion.

- **Flat bones,** such as the skull, sternum, scapulae, and pelvic bones, shield soft tissues.

- **Irregular bones,** such as the mandible (jawbone) and vertebrae, come in a variety of shapes and sizes suited for attachment to muscles, tendons, and ligaments. Irregular bones include seed-shaped bones found in joints such as the *patella,* or kneecap.

Unfortunately for students of bone structures, there's no easy way to memorize them. So brace yourself for a rapid summary of what your textbook probably goes into in much greater detail.

Compact bone is composed of dense layers of *lamellae* surrounding *lacunae,* small concavities containing osteocytes, arranged in concentric circles called *Haversian systems* (structural units also referred to in short as *osteons*), each of which has a central, microscopic *Haversian canal.* A perpendicular system of canals, called *Volkmann's canals,* penetrate and cross between the Haversian systems. This network ensures circulation into even the hardest bone structure. Compact bone tissue is thick in the shaft and tapers to paper thinness at the ends of the bones. The bulbous ends of each long bone, known as the *epiphyses* (or singularly as an *epiphysis*), are made up of spongy, or *cancellous,* bone tissue covered by a thin layer of compact bone. The *diaphysis,* or shaft, contains the *medullary cavity* (*medullary* derives from Latin for *marrow*) containing blood cell–producing marrow supplied by a medullary nutrient artery. A membrane called the *periosteum* covers the outer bone, except for the epiphyseal surfaces, to provide nutrients and oxygen, remove waste, and connect with ligaments and tendons.

Bones grow through the cellular activities of *osteoblasts* on the surface of the bone, which differentiate into layers of mature bone cells called *osteocytes. Osteoclasts* are cells that function in the developing fetus to absorb cartilage as ossification occurs and function in adult bone to break down and remove spent bone tissue.

There are two types of *ossification,* which is the process by which softer tissues harden into bone. Both types rely on a peptide hormone produced by the thyroid gland, *calcitonin,* which regulates metabolism of calcium, the body's most abundant mineral. The two types of ossification are

✔ **Endochondral or intracartilaginous ossification:** Occurs when mineral salts, particularly calcium and phosphorus, calcify along the scaffolding of cartilage formed in the developing fetus beginning about the fifth week after conception. This process, known as *calcification,* takes place in the presence of vitamin D and a hormone from the parathyroid gland. The absence of any one of these substances causes a child to have soft bone, called *rickets.* Next, the blood supply entering the cartilage brings osteoblasts that attach themselves to the cartilage. As the primary center of ossification, the diaphysis of the long bone is the first to form spongy bone tissue along the cartilage, followed by the epiphyses, which form the secondary centers of ossification and are separated from the diaphysis by a layer of uncalcified cartilage called the *epiphyseal plate,* where all growth in bone length occurs. Compact bone tissue covering the bone's surface is produced by osteoblasts in the inner layer of the periosteum, producing growth in diameter.

✔ **Intramembranous ossification:** Occurs not along cartilage but instead along a template of membrane, as the name implies, primarily in compact flat bones of the skull that don't have Haversian systems. The skull and mandible (lower jaw) of the fetus are first laid down as a membrane. Osteoblasts entering with the blood supply attach to the membrane, ossifying from the center of the bone outward. The edges of the skull's bones don't completely ossify to allow for molding of the head during birth. Instead, six soft spots, or *fontanels,* are formed: one frontal or anterior, two sphenoidal or anterolateral, two mastoidal or posterolateral, and one occipital or posterior.

Once formed, bone is surrounded by the periosteum, which has both a vascular layer (the Latin word for "vessel" is *vasculum*) and an inner layer that contains the osteoblasts needed for bone growth and repair. A penetrating matrix of connective tissue, called *Sharpey's fibers,* connects the periosteum to the bone. Inside the bone, a thin membrane called the *endosteum* (from the Greek *endon,* meaning "within," and the Greek *osteon* meaning "bone") lines the medullary cavity.

Following are the basic terms used to identify bone landmarks or surface features:

✔ **Process:** A broad designation for any prominence or prolongation

✔ **Spine:** An abrupt or pointed projection

✔ **Trochanter:** A large, usually blunt process

✔ **Tubercle:** A smaller, rounded eminence

✔ **Tuberosity:** A large, often rough eminence

✔ **Crest:** A prominent ridge

✔ **Head:** A large, rounded articular end of a bone, often set off from the shaft by a neck (*articular* is an adjective describing areas related to movement between bones)

✔ **Condyle:** An oval articular prominence of a bone

✔ **Facet:** A smooth, flat, or nearly flat articulating surface

✔ **Fossa:** A deeper depression

> ✔ **Sulcus:** A groove
>
> ✔ **Foramen:** A hole
>
> ✔ **Meatus:** A canal or opening to a canal

5. Why do bones break more easily in older bodies than in younger ones?

   a. They wear away with time and use.

   b. Minerals in the bones increase over time, making them more brittle.

   c. Older bones have larger gaps between supportive structures.

   d. An older person's center of gravity is higher, so bones impact the ground harder in a fall.

6. Why would intramembranous ossification not be observed in a 50-year-old?

   a. The joints would not tolerate it.

   b. Broken bones heel more slowly in adults.

   c. It only occurs in fetuses and very young children.

   d. Muscles don't attach to those areas.

7. What does calcitonin do?

   a. It regulates metabolism of calcium.

   b. It controls the development of Haversian systems.

   c. It influences the formation of cardiac tissue.

   d. It encourages the proper layering of osteoclasts.

8. What does the periosteum do?

   a. It ensures circulation into even the hardest bone structures.

   b. It covers bones to provide nutrients and oxygen, remove waste, and connect with ligaments and tendons.

   c. It controls calcification over time.

   d. It allows flexibility in infant bone structures that is needed for proper development.

9. Where are Sharpey's fibers?

   a. Inside Haversian canals

   b. Beside most sphenoids

   c. Adjacent to both the periosteum and the bone, connecting the two

   d. Behind any lingering mastoid tissue

10. What is inside the medullary cavity?

   a. Bone marrow

   b. Epiphyses

   c. Volkmann's canals

   d. Osteoclasts

11. Why does a baby have fontanels?

   a. To regulate temperature changes immediately after birth

   b. For the proper development of dental structures

   c. To direct the growth of long bones

   d. To allow for molding of the head during birth

12. Volkmann's canals

   a. are found in compact bone only.

   b. contain the nutrient artery.

   c. pass through the epiphysis.

   d. supply blood to articulating cartilage.

13. Which type of bones forms the weight-bearing part of the skeleton?

   a. Flat bones

   b. Irregular bones

   c. Long bones

   d. Short bones

14. What causes rickets?

   a. Hyperactivity in the thyroid

   b. Too much calcitonin

   c. A lack of citrus acid

   d. Insufficient vitamin D

15.–25. Fill in the blanks to complete the following sentences:

Bones are first laid down as 15. _____ during the fifth week after conception. Development of the bone begins with 16._____, the depositing of calcium and phosphorus. Next, the blood supply brings 17. _____ that attach themselves to the cartilage. Ossification in long bones begins in the 18. _____ of the long bone and moves toward the 19. _____ of the bone. These areas remain separated by a layer of uncalcified cartilage called the 20. _____.

Another very large cell that enters with the blood supply is the 21. _____, which helps absorb the original material as ossification occurs. Later it helps absorb bone tissue from the center of the long bone's shaft, forming the 22. _____ cavity. After ossification, the spaces that were formed join together to form 23. _____ systems, which contain the blood vessels, lymphatic vessels, and nerves. Unlike bones in the rest of the body, those of the skull and mandible (lower jaw) are first laid down as 24. _____. In the skull, the edges of the bone don't ossify in the fetus but remain membranous and form 25. _____.

26.–34. Classify the following bones by shape. Each classification may be used more than once.

26. _____ Vertebrae of the vertebral column

27. _____ Femur in thigh

28. _____ Sternum

29. _____ Tarsals in ankle

30. _____ Humerus in upper arm

31. _____ Phalanges in fingers and toes

32. _____ Scapulae of shoulder

33. _____ Kneecap

34. _____ Carpals in wrist

a. Flat bone

b. Irregular bone

c. Long bone

d. Short bone

35.–47. Match the description with the bone landmarks or surface features.

35. _____ An abrupt or pointed projection

36. _____ A large, usually blunt process

37. _____ A designation for any prominence or prolongation

38. _____ A large, often rough eminence

39. _____ A prominent ridge

40. _____ A large, rounded articular end of a bone; often set off from the shaft by the neck

41. _____ An oval articular prominence of a bone

42. _____ A smooth, flat, or nearly flat articulating surface

43. _____ A deeper depression

44. _____ A groove

45. _____ A hole

46. _____ A canal or opening to a canal

47. _____ A smaller, rounded eminence

a. Condyle

b. Crest

c. Facet

d. Foramen

e. Fossa

f. Head

g. Meatus

h. Process

i. Spine

j. Sulcus

k. Trochanter

l. Tubercle

m. Tuberosity

48.–56. Use the terms that follow to identify the regions and structures of the long bone shown in Figure 5-1.

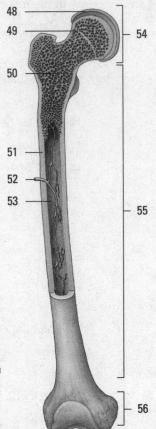

Figure 5-1:
The long
bone.

Illustration by Imagineering Media Services Inc.

    a. Diaphysis

    b. Medullary cavity

    c. Distal epiphysis

    d. Spongy bone tissue

    e. Medullary or nutrient artery

    f. Proximal epiphysis

    g. Red bone marrow

    h. Articular cartilage

    i. Compact bone tissue

The Axial Skeleton: Keeping It All in Line

Just as the Earth rotates around its axis, the axial skeleton lies along the midline, or center, of the body. Think of your spinal column and the bones that connect directly to it — the rib (thoracic) cage and the skull. The tiny *hyoid* bone, which lies just above your larynx, or voice box, also is considered part of the axial skeleton, although it's the only bone in the entire body that doesn't connect, or articulate, with any other bone. It's also known as the *tongue bone* because the tongue's muscles attach to it. There are a total of 80 named bones in the axial skeleton, which supports the head and trunk of the body and serves as an anchor for the pelvic girdle.

Making a hard head harder

Of the 80 named bones in the axial skeleton, 29 are in (or very near) the skull. In addition to the hyoid bone, 8 bones form the cranium to house and protect the brain, 14 form the face, and 6 bones make it possible for you to hear.

Fortunately for the cramming student, most of the bones in the skull come in pairs. In the cranium there's just one of each of the following:

- ✔ *Frontal* bone (forehead)

- ✔ *Occipital* bone (back and base of the skull) containing occipital condyles, which articulate with the atlas of the vertebral column

- ✔ *Ethmoid* bone (made of several plates, or sections, between the eye orbits in the nasal cavity)

- ✔ *Sphenoid* bone (a butterfly-shaped structure that forms the floor of the cranial cavity)

But there are two *temporal* bones each housing: (1) the hearing organs in the *auditory meatus;* (2) a styloid process, a long pointed process for anchoring muscles; (3) a mandibular fossa articulating with the condyle of mandible; and (4) a zygomatic process that projects anteriorly joining the zygomatic bone forming the cheek prominence. There are two *parietal* bones (roof and sides of the skull). These bones are attached along *sutures* called the following:

- ✔ *Coronal* (located at the top of the skull between the two parietal bones)

- ✔ *Squamosal* (located on the sides of the head surrounding the temporal bone)

- ✔ *Sagittal* (along the midline atop the skull located between the two parietal bones)

- ✔ *Lambdoidal* (forming an upside-down V — the shape of the Greek letter lambda — on the back of the skull)

In the face, there is only one *mandible* (jawbone) and one *vomer* dividing the nostrils, but there are two each of *maxillary* (upper jaw), *zygomatic* (cheekbone), *nasal, lacrimal* (a small bone in the eye socket), *palatine* (which makes up part of the eye socket, nasal cavity, and roof of the mouth), and *inferior nasal concha,* or turbinated, bones. The mandible has two lateral condyles that articulate in the mandibular fossa of the temporal bone. (The maxillary bones are also called, simply, the *maxilla.*) Inside the ear, there are two each of three ossicles, or bonelets, which also happen to be the smallest bones in the human body: the *malleus, incus,* and *stapes.*

The floor of the cranial cavity contains several openings, or *foramina* (the singular is *foramen*), that allow various nerves and vessels to connect to the brain.

✔ The holes in the ethmoid bone's *cribriform plate* are *olfactory foramina* that allow olfactory — or sense of smell — receptors to pass through to the brain. A process called the *crista galli* extends into the brain cavity for the attachment of the *meninges* of the brain.

✔ A large hole in the occipital bone called the *foramen magnum* allows the spinal cord to connect with the brain.

✔ The sphenoid bone is riddled with foramina.

  • The *optic foramen* allows passage of the optic nerves.

  • The *jugular foramen* allows passage of the jugular vein and several cranial nerves.

  • The *foramen rotundum* allows passage of the *trigeminal nerve,* which is the chief sensory nerve to the face and controls the motor functions of chewing.

  • The *foramen ovale* allows passage of the nerves controlling the tongue, among other things.

  • The *foramen spinosum* allows passage of the middle *meningeal* artery, which supplies blood to various parts of the brain.

  • The sphenoid bone also features the *sella turcica,* or Turk's saddle, that cradles the pituitary gland and forms part of the *foramen lacerum,* through which pass several key components of the autonomic nervous system.

Encased within the frontal, sphenoid, ethmoid, and maxillary bones of the skull are several air-filled, mucous-lined cavities called *paranasal sinuses.* Although you may think their primary function is to drive you crazy with pressure and infections, the sinuses actually lighten the skull's weight. They make it easier to hold your head up high; they warm and humidify inhaled air; and they act as resonance chambers to prolong and intensify the reverberations of your voice. *Mastoid sinuses* drain into the middle ear (hence the earache referred to as *mastoiditis*). *Maxillary sinuses* are flanked by the bones of the maxilla, the upper jaw. The *paranasal sinuses* are the ones that drain into the nose and cause so much trouble when you cry or have a cold.

Putting your backbones into it

REMEMBER

The axial skeleton also consists of 33 bones in the vertebral column, laid out in four distinct curvatures, or areas.

✔ **The cervical, or neck, curvature** has seven vertebrae, with the *atlas* and *axis* bones positioned in the first and second spots, respectively, forming a joint connecting the skull and vertebral column. The axis, or second cervical vertebra, contains the *dens,* or *odontoid process.* The atlas rotates around the dens, turning the head. (In a sense, the atlas bone holds the world of the head on its shoulders, as the Greek god Atlas held the Earth.)

✔ **The thoracic, or chest, curvature** has 12 vertebrae that articulate with the ribs, most of which attach to the sternum anteriorly by costal cartilage, forming the rib cage that protects the heart and lungs.

✔ **The lumbar, or small of the back, curvature** contains five vertebrae and carries most of the weight of the body, which means that it generally suffers the most stress.

✔ **The pelvic curvature** includes the five fused vertebrae of the *sacrum* anchoring the pelvic girdle by the sacroiliac articular joint and four fused vertebrae of the *coccyx*, or tailbone.

The spinal cord extends down the center of the vertebrae only from the base of the brain to the uppermost lumbar vertebrae.

Each vertebra, except the atlas, consists of a body and a vertebral arch, which features a long dorsal projection called a *spinous process* that provides a point of attachment for muscles and ligaments. On either side of this are the *laminae,* broad plates of bone on the posterior surface that form a bony covering over the spinal canal. The laminae attach to the two *transverse processes,* which in turn are attached to the body of the vertebra by regions called the *pedicles.* The vertebrae align to form a large opening, called the *vertebral foramen,* allowing the passage of the spinal cord. Laterally, between the vertebrae, are openings called the *intervertebral foramina* that allow the spinal nerves to exit the vertebral column. The vertebra have superior articulating facets that articulate with inferior articulating facets of the adjacent vertebra, increasing the rigidity of the column, making the backbone more stable. Fibrocartilage discs located between the vertebrae act as shock absorbers. Openings in the transverse process, called the *transverse foramina,* allow large vessels and nerves ascending the neck to reach the brain.

Connecting to the vertebral column are the 12 pairs of ribs that make up the thoracic cage. All 12 pairs attach to the thoracic vertebrae, but the first 7 pairs attach to the *sternum,* or breastbone, by *costal cartilage;* they're called *true ribs.* Pairs 8, 9, and 10 attach to the cartilage of the seventh pair, which is why they're called *false ribs.* The last two pairs aren't attached in front at all, so they're called *floating ribs.*

The sternum has three parts:

✔ **Manubrium:** The superior region that articulates with the clavicle, at the clavicular notch, and the first two pairs of ribs located up top, where you can feel a jugular notch in your chest in line with your *clavicles,* or collarbones.

✔ **Body:** The middle part of the sternum forms the bulk of the breastbone and has notches on the sides where it articulates with the third through seventh pairs of ribs.

✔ **Xiphoid process:** The lowest part of the sternum is an attachment point for the diaphragm and some abdominal muscles. (Interesting fact: Emergency medical technicians learn to administer CPR at least three finger widths above the xiphoid.)

57.–71. Use the terms that follow to identify the bones, sutures, and landmarks of the skull shown in Figure 5-2.

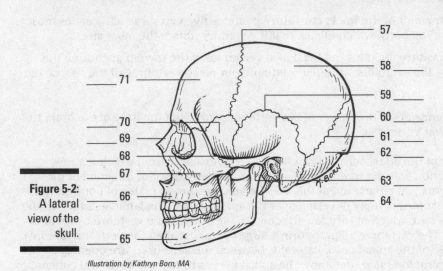

Figure 5-2:
A lateral
view of the
skull.

Illustration by Kathryn Born, MA

a. Temporal bone

b. Sphenoid bone

c. Nasal bone

d. Styloid process

e. Frontal bone

f. Squamosal suture

g. Maxilla

h. Parietal bone

i. Lambdoidal suture

j. Occipital bone

k. Zygomatic bone

l. Coronal suture

m. Mandibular condyle

n. Zygomatic process

o. Mandible

72.–75. Match the skull bones with their connecting sutures.

72. _____ Frontal and parietals   **a.** Squamosal suture

73. _____ Occipital and parietals   **b.** Lambdoidal suture

74. _____ Parietal and parietal   **c.** Sagittal suture

75. _____ Temporal and parietal   **d.** Coronal suture

76.–82. Use the terms that follow to identify the bones and landmarks of the skull shown in Figure 5-3.

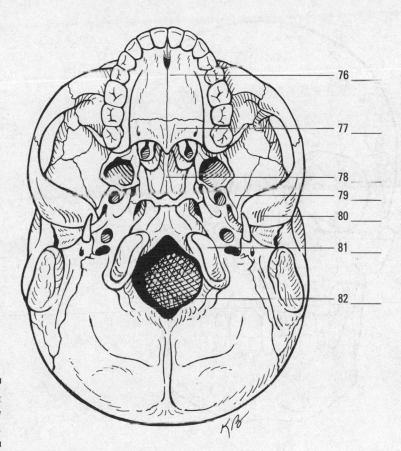

76 _____

77 _____

78 _____

79 _____

80 _____

81 _____

82 _____

Figure 5-3:
Inferior view
of the skull.

Illustration by Kathryn Born, MA

    a. Vomer

    b. Mandibular fossa

    c. Foramen magnum

    d. Palatine bone

    e. Occipital condyle

    f. Maxilla

    g. Sphenoid bone

83.–88. Use the terms that follow to identify the bones and landmarks of the skull shown in Figure 5-4.

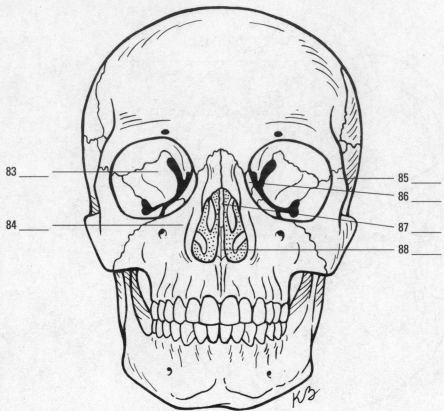

Figure 5-4:
Frontal view
of the skull.

Illustration by Kathryn Born, MA

a. Zygomatic bone

b. Vomer

c. Ethmoid bone

d. Sphenoid bone

e. Lacrimal bone

f. Maxilla

89.–105. Use the terms that follow to identify the bones and landmarks of the cranial cavity shown in Figure 5-5.

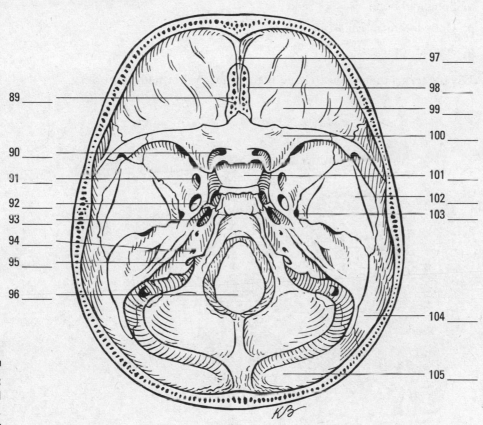

Figure 5-5:
Cranial
cavity.

Illustration by Kathryn Born, MA

a. Internal auditory (acoustic) meatus

b. Parietal bone

c. Foramen ovale

d. Frontal bone

e. Foramen lacerum

f. Cribriform plate

g. Sella turcica

h. Temporal bone

i. Foramen spinosum

j. Occipital bone

k. Olfactory foramina

l. Foramen rotundum

m. Foramen magnum

n. Crista galli

o. Sphenoid bone

p. Optic foramen (canal)

q. Jugular foramen

106.–109. Use the terms that follow to identify the sinuses shown in Figure 5-6.

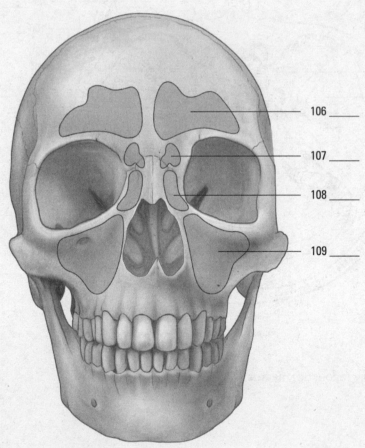

106 _____

107 _____

108 _____

109 _____

Figure 5-6:
Sinus view
of the skull.

Illustration by Imagineering Media Services Inc.

a. Sphenoid sinus

b. Frontal sinus

c. Maxillary sinus

d. Ethmoid sinus

110.–118. Use the terms that follow to identify the regions, structures, and landmarks of the vertebral column shown in Figure 5-7.

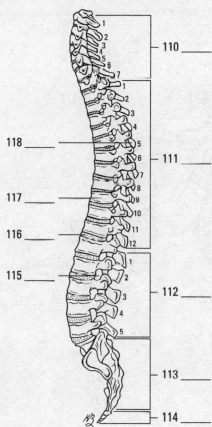

110 _____

118 _____

111 _____

117 _____

116 _____

115 _____

112 _____

113 _____

114 _____

Figure 5-7:
Vertebral
column.

Illustration by Kathryn Born, MA

a. Coccyx

b. Intervertebral foramen

c. Thoracic vertebrae or curvature

d. Sacrum

e. A vertebra

f. Cervical vertebrae or curvature

g. Lumbar vertebrae or curvature

h. A spinous process

i. Intervertebral disc

119.–127. Use the terms that follow to identify the landmarks of the vertebra shown in Figure 5-8.

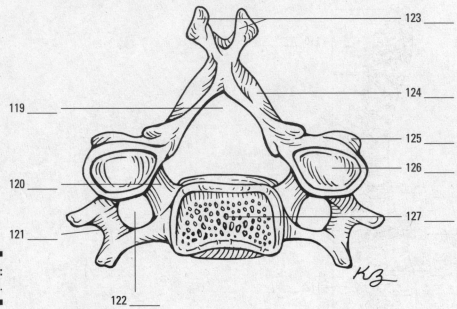

Figure 5-8: Vertebra.

Illustration by Kathryn Born, MA

a. Vertebral foramen

b. Lamina

c. Transverse process

d. Spinous process (bifid)

e. Superior articulating facet

f. Body

g. Transverse foramen

h. Inferior articulating facet

i. Pedicle

128.–133. Use the terms that follow to identify the regions, landmarks, and structures of the thoracic cage shown in Figure 5-9.

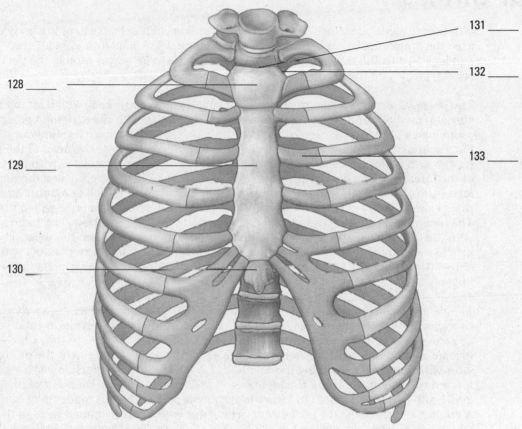

128 _____

129 _____

130 _____

131 _____

132 _____

133 _____

Figure 5-9:
Thoracic
cage.

Illustration by Imagineering Media Services Inc.

    a. Xiphoid process

    b. Jugular notch

    c. Body

    d. Costal cartilage

    e. Clavicular notch

    f. Manubrium

134.–139. Fill in the blanks to complete the following sentences:

The organs protected by the thoracic cage include the 134. _____ and the 135. _____. The first seven pairs of ribs attach to the sternum by the costal cartilage and are called 136. _____ ribs. Pairs 8 through 10 attach to the costal cartilage of the seventh pair and not directly to the sternum, so they're called 137. _____ ribs. The last two pairs, 11 and 12, are unattached anteriorly, so they're called 138. _____ ribs. There's one bone in the entire skeleton that doesn't articulate with any other bones but nonetheless is considered part of the axial skeleton. It's called the 139. _____ bone.

The Skeleton: Reaching beyond Our Girdles

Whereas the axial skeleton (described earlier in this chapter) lies along the body's central axis, the appendicular skeleton's 126 bones include those in all four appendages — arms and legs — plus the two primary girdles to which the appendages attach: the pectoral (chest) girdle and the pelvic (hip) girdle.

The pectoral girdle is made up of a pair of clavicles, or collarbones, which attach to the sternum medially and to the scapula laterally articulating with the *acromion process,* a bony prominence at the top of each of the pair of scapulae, better known as shoulder blades. Each scapula has a depression in it called the *glenoid fossa* where the head of the *humerus* (upper arm bone) is attached. The *capitulum* of the humerus articulates with the trochlea of the forearm's long *ulna* bone to form the elbow joint. The process called the *olecranon* forms the elbow and is also referred to as the funny bone, although banging it into something usually feels anything but funny. The forearm also contains a bone called the *radius.* The biceps muscle attaches to the radial tuberosity, flexing the elbow. The radius, together with the ulna, articulates with the eight small *carpal* bones that form the wrist. The carpals articulate with the five *metacarpals* that form the hand, which in turn connect with the *phalanges* (finger bones), which are found as a pair in the thumb and as triplets in each of the fingers.

The pelvic girdle consists of two hipbones, called *os coxae,* as well as the *sacrum* and *coccyx,* more commonly referred to as the tailbone. During early developmental years, the os coxa consists of three separate bones — the *ilium,* the *ischium,* and the *pubis* — that later fuse into one bone sometime between the ages of 16 and 20. Posteriorly, the os coxa articulates with the sacrum, forming the *sacroiliac joint,* the source of much lower back pain; it's formed by the connection of the hipbones at the sacrum. Toward the front of the pelvic girdle, the two os coxae join to form the *symphysis pubis,* which is made up of fibrocartilage. A cuplike socket called the *acetabulum* articulates with the ball-shaped head of the *femur* (thigh bone). The femur is the longest bone in the body. The femur articulates with the *tibia* (shin bone) at the knee, which is covered by the *patella* (kneecap). Also inside each lower leg is the *fibula* bone, which joins with the tibia to connect with the seven *tarsal* bones that make up the ankle. These are the calcaneus (heel bone), talus, navicular, cuboid, lateral cuneiform, intermediate cuneiform, and medial curneiform. The largest of these bones is the calcaneus. The tarsals join with the five *metatarsals* that form the foot, which in turn connect to the *phalanges* of the toes — a pair of phalanges in the big toe and triplets in each of the other toes.

140.–159. Use the terms that follow to identify the bones and structures of the appendicular skeleton shown in Figure 5-10.

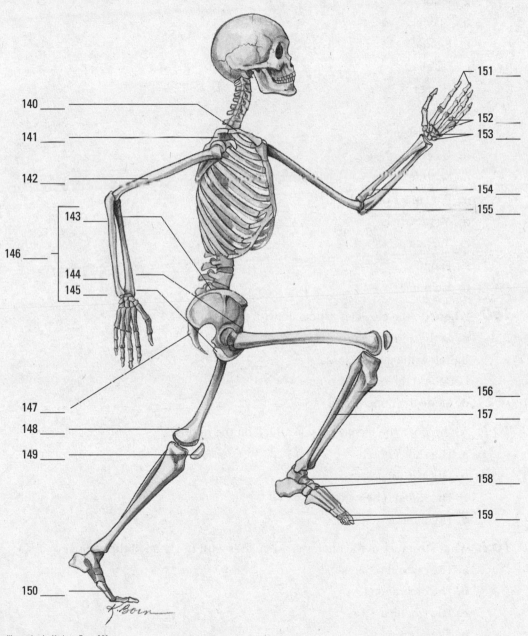

140 ____
141 ____
142 ____
143 ____
144 ____
145 ____
146 ____
147 ____
148 ____
149 ____
150 ____
151 ____
152 ____
153 ____
154 ____
155 ____
156 ____
157 ____
158 ____
159 ____

Figure 5-10:
The appendicular skeleton.

Illustration by Kathryn Born, MA

a. Tibia

b. Ulna

c. Scapula

d. Metatarsals

e. Carpals

f. Phalanges of the foot

g. Ilium

h. Clavicle

i. Fibula

j. Patella

k. Humerus

l. Pubis

m. Radius

n. Os coxa

o. Sacrum

p. Phalanges of the hand

q. Metacarpals

r. Tarsals

s. Femur

t. Ischium

160. Where do the clavicles articulate with the scapulae?

a. At the acromion process

b. Below the glenoid fossa

c. Behind the coracoid process

d. Along the upper spine

161. Where does the biceps muscle attach on the radius?

a. The radial notch

b. The styloid process

c. The radial tuberosity

d. The ulnar notch

162. What structure of the humerus articulates with the head of the radius?

a. The deltoid tuberosity

b. The olecranon fossa

c. The trochlea

d. The capitulum

163. The patella is what kind of bone?

a. Wormian

b. Sesamoid

c. Pisiform

d. Hallux

164. Which of these bones is not part of the pelvic girdle?

    a. Ilium

    b. Lumbar vertebrae

    c. Sacrum

    d. Ischium

165. What is the formal term for the prominence commonly referred to as the elbow?

    a. Olecranon process

    b. Trochlear notch

    c. Radial notch

    d. Coronoid process

166. Where does the ulna articulate with the humerus?

    a. The deltoid tuberosity

    b. The greater tubercle

    c. The capitulum

    d. The trochlea

167. What is the formal term for the socket at the head of the femur?

    a. Obturator foramen

    b. Acetabulum

    c. Ischial tuberosity

    d. Greater sciatic notch

168. Which is largest and strongest tarsal bone?

    a. The talus

    b. The cuboid

    c. The cavicular

    d. The calcaneus

169. Where are people who complain about their sacroiliac experiencing pain?

    a. Lower back

    b. Neck

    c. Feet

    d. Hands

Arthrology: Articulating the Joints

Arthrology, which stems from the ancient Greek word *arthros* (meaning "jointed"), is the study of those structures that hold bones together, allowing them to move to varying degrees — or fixing them in place — depending on the design and function of the joint. The term *articulation,* or *joint,* applies to any union of bones, whether it moves freely or not at all.

Inside some joints, such as knees and elbows, are fluid-filled sacs called *bursae* that help reduce friction between tendons and bones; inflammation in these sacs is called *bursitis.* Syndesmosis joints are stabilized by connective tissue called *ligaments,* or *interosseus membranes,* that range from bundles of collagenous fibers that restrict movement and hold a joint in place to elastic fibers that can repeatedly stretch and return to their original shapes. Examples are

- ✔ The shoulder joint, or corticohumeral ligament that extends from the coracoid process of the scapula to the greater tubercle of the humerus
- ✔ The pubofemoral ligament that extends from the pubic bone to the femur
- ✔ The knee joint (oblique popliteal ligament), where the tendon of the semimembranous muscle expands to cross the posterior of the knee joint

The three types of joints are as follows:

- ✔ **Fibrous:** Fibrous tissue rigidly joins the bones in a form of articulation called *synarthrosis,* which is characterized by no movement at all. The sutures of the skull are fibrous joints. Any slight movement in a joint depends on the length of the fibers uniting the bones.

- ✔ **Cartilaginous:** This type of joint is found in two forms:

  - • *Synchondrosis articulation* involves *hyaline* (rigid) cartilage that allows no movement. Once bone growth ends, the joint becomes ossified and immobile. The most common example is the epiphyseal plate of the long bone. Other examples are the joint between the ribs, costal cartilage, and sternum.

  - • *Symphysis joints* occur where fibrocartilage fuses bones in such a way that pressure can cause slight movement, called *amphiarthrosis.* Examples include the intervertebral discs and the symphysis pubis.

- ✔ **Synovial:** Also known as *diarthrosis,* or freely moving, joints, this type of articulation involves a synovial cavity, which contains articular fluid secreted from the synovial membrane to lubricate the opposing surfaces of bone covered by smooth articulating cartilage. The synovial membrane is covered by a fibrous joint capsule layer that's continuous with the periosteum of the bone. Ligaments surrounding the joint strengthen the capsule and hold the bones in place, preventing dislocation. In some synovial joints, such as the knee, fibrocartilage pads called *menisci* (singular: *meniscus*) develop in the cavity, dividing it into two parts. In the knees, these menisci stabilize the joint and act as shock absorbers.

There are six classifications of moveable, or synovial, joints:

- ✔ **Gliding:** Curved or flat surfaces slide against one another, such as between the carpal bones in the wrist or between the tarsal bones in the ankle.

✔ **Hinge:** A convex surface joints with a concave surface, allowing right-angle motions in one plane, such as elbows, knees, and joints between the finger bones.

✔ **Pivot (or rotary):** One bone pivots or rotates around a stationary bone, such as the atlas rotating around the odontoid process of the axis at the top of the vertebral column.

✔ **Condyloid:** The oval head of one bone fits into a shallow depression in another, to allow for five movements: flexion, extension, abduction, adduction, and circumduction. Examples are the carpal-metacarpal joint at the wrist and the tarsal-metatarsal joint at the ankle.

✔ **Saddle:** Each of the adjoining bones is shaped like a saddle (the technical term is *reciprocally concavo-convex*). The saddle joints resemble condyloid joints but have even greater freedom of movement. An example is the carpometacarpal joint of the thumb.

✔ **Ball-and-socket:** The round head of one bone fits into a cuplike cavity in the other bone, allowing movement in many directions so long as the bones are neither pulled apart nor forced together, such as the shoulder joint between the humerus and scapula and the hip joints between the femur and the os coxa.

The following are the types of joint movement:

✔ **Flexion:** A decrease in the angle between two bones

✔ **Extension:** An increase in the angle between two bones

✔ **Abduction:** Movement away from the midline of the body

✔ **Adduction:** Movement toward the midline of the body

✔ **Rotation:** Turning around an axis

✔ **Pronation:** Downward or palm downward

✔ **Supination:** Upward or palm upward

✔ **Eversion:** Turning of the sole of the foot outward

✔ **Inversion:** Turning of the sole of the foot inward

✔ **Circumduction:** Forming a cone with the arm or leg

170.–175. Match the articulations with their joint types. Some joint types may be used more than once.

170. _____ Sutures of the skull

171. _____ Fluid-filled cavity

172. _____ Knee joint

173. _____ Symphysis pubis

174. _____ Epiphyseal plate

175. _____ Intervertebral discs

a. Fibrous joint

b. Cartilaginous joint-synchondrosis

c. Cartilaginous joint-symphysis

d. Synovial joint

176.—180. Use the terms that follow to identify the structures that form the synovial joint shown in Figure 5-11.

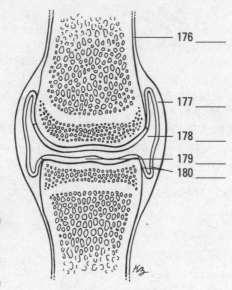

176 _____

177 _____

178 _____

179 _____

180 _____

Figure 5-11:
A synovial
joint.

Illustration by Kathryn Born, MA

a. Synovial (joint) cavity

b. Periosteum

c. Synovial membrane

d. Articular cartilage

e. Fibrous capsule

181. What is the formal term for an immovable joint?

a. An amphiarthrosis

b. A synarthrosis

c. A diarthrosis

d. A synchondrosis

182. What is the formal term for a freely moving joint?

a. An amphiarthrosis

b. A synarthrosis

c. A diarthrosis

d. A synchondrosis

183. What material or structure allows free movement in a joint?

    a. The bursa

    b. The periosteum

    c. Synovial fluid

    d. Bone marrow

184. Choose an example of a ball-and-socket joint.

    a. The symphysis pubis

    b. The hip

    c. The ankle

    d. The elbow

185. Choose an example of a pivotal joint.

    a. Between the radius and the ulna

    b. The interphalangeal joints

    c. Between the mandible and the temporal bone

    d. Between the tibia and the fibula

186. Choose an example of a saddle joint.

    a. The radius and the carpals

    b. The carpometacarpal joint of the thumb

    c. The occipital condyles and the atlas

    d. The metatarsophalanges joint

187. Choose an example of a shoulder joint ligament.

    a. Coracohumeral ligament

    b. Popliteal ligament

    c. Ischiofemoral ligament

    d. Ligamentum arteriosum

188. Choose an example of a knee joint ligament.

    a. Iliofemoral ligament

    b. Coracohumeral ligament

    c. Oblique popliteal ligament

    d. Annular ligament

189. Choose an example of a hip joint ligament.

    a. Subscapularis ligament

    b. Pubofemoral ligament

    c. Glenohumeral ligament

    d. Coracohumeral ligament

190. What is the structure in the knee that divides the synovial joint into two separate compartments?

    a. The bursa

    b. Joint fat

    c. The tendon sheath

    d. The meniscus, or articular disc

191.–200. Match the type of joint movement with its description.

191. _____ Flexion		**a.** Upward or palm upward
192. _____ Extension		**b.** Decrease in the angle between two bones
193. _____ Abduction		**c.** Turning of the sole of the foot inward
194. _____ Adduction		**d.** Downward or palm downward
195. _____ Rotation		**e.** Increase in the angle between two bones
196. _____ Pronation		**f.** Turning of the sole of the foot outward
197. _____ Supination		**g.** Movement away from the midline of the body
198. _____ Eversion		**h.** Forming a cone with the arm or leg
199. _____ Inversion		**i.** Turning around an axis
200. _____ Circumduction		**j.** Movement toward the midline of the body

Answers to Questions on the Skeleton

The following are answers to the practice questions presented in this chapter.

1 What is happening during hematopoiesis? **b. Bone marrow is forming red blood cells.** Most of the other answers are nonsense.

2 Why do the bones contain minerals like phosphorous, calcium, and magnesium? **c. To ensure availability of these minerals, even if insufficient amounts of them are consumed.** The bones act as a mineral bank for the entire body.

3 In what way does the skeleton make locomotion possible? **b. Muscles use the skeleton as a scaffold against which they can contract.** You'd be a motionless blob without a skeleton to support coordinated movement, and, besides, the rest of the answers are utter nonsense.

4 The curvatures in some bone structures serve the following purpose: **a. Support the body.** And now you know why your back aches most where it curves.

5 Why do bones break more easily in older bodies than in younger ones? **b. Minerals in the bones increase over time, making them more brittle.** The correct answer becomes obvious through a process of elimination.

6 Why would intramembranous ossification not be observed in a 50-year-old? **c. It only occurs in fetuses and very young children.** This type of ossification follows a membrane template, such as over the soft fontanels of a baby's skull.

7 What does calcitonin do? **a. It regulates metabolism of calcium.** Once you remember that calcitonin comes from the metabolism's pacemaker, the thyroid gland, you can discern the right answer.

8 What does the periosteum do? **b. It covers bones to provide nutrients and oxygen, remove waste, and connect with ligaments and tendons.** Back to Greek again: *peri* means "around" and *osteon* means "bone," so the periosteum is "around the bone."

9 Where are Sharpey's fibers? **c. Adjacent to both the periosteum and the bone, connecting the two.** Described by anatomist William Sharpey in 1846, these are also called *perforating fibers*.

10 What is inside the medullary cavity? **a. Bone marrow.** Usually you'll find yellow marrow, although in infants red marrow also is present.

11 Why does a baby have fontanels? **d. To allow for molding of the head during birth.** These six separate floating plates are why you can see a bald baby's pulse throbbing on the top of its head.

12 Volkmann's canals **a. are found in compact bone only.** Ironically, anatomist Alfred Wilhelm Volkmann was most noted for his observations of the physiology of the nervous system, not bones.

13 Which type of bones forms the weight-bearing part of the skeleton? **c. Long bones.** After all, they are the largest and most robust of the body's bones.

14 What causes rickets? **d. Insufficient vitamin D.** Rickets is a condition of soft bones in children whose skeletons didn't have what was needed for them to ossify properly.

15–25 Fill in the blanks to complete the following sentences:

Bones are first laid down as **15. cartilage** during the fifth week after conception. Development of the bone begins with **16. calcification,** the depositing of calcium and phosphorus. Next, the blood supply entering the cartilage brings **17. osteoblasts** that attach themselves to the cartilage. Ossification in long bones begins in the **18. diaphysis** of the long bone and moves toward the **19. epiphysis** of the bone. The epiphyseal and diaphyseal areas remain separated by a layer of uncalcified cartilage called the **20. epiphyseal plate.**

Another very large cell that enters with the blood supply is the **21. osteoclast,** which helps absorb the cartilage as ossification occurs. Later it helps absorb bone tissue from the center of the long bone's shaft, forming the **22. medullary or marrow** cavity. After ossification, the spaces that were formed by the osteoclasts join together to form **23. Haversian canal** systems, which contain the blood vessels, lymphatic vessels, and nerves. Unlike bones in the rest of the body, those of the skull and mandible (lower jaw) are first laid down as **24. membrane.** In the skull, the edges of the bone don't ossify in the fetus but remain membranous and form **25. fontanels.**

26 Vertebrae of the vertebral column: **b. Irregular bone**

27 Femur in thigh: **c. Long bone**

28 Sternum: **a. Flat bone**

29 Tarsals in ankle: **d. Short bone**

30 Humerus in upper arm: **c. Long bone**

31 Phalanges in fingers and toes: **c. Long bone**

32 Scapulae of shoulder: **a. Flat bone**

33 Kneecap: **b. Irregular bone**

34 Carpals in wrist: **d. Short bone**

35 An abrupt or pointed projection: **i. Spine**

36 A large, usually blunt process: **k. Trochanter**

37 A designation for any prominence or prolongation: **h. Process**

38 A large, often rough eminence: **m. Tuberosity**

39 A prominent ridge: **b. Crest**

40 A large, rounded articular end of a bone, often set off from the shaft by the neck: **f. Head**

41 An oval articular prominence of a bone: **a. Condyle**

42 A smooth, flat, or nearly flat articulating surface: **c. Facet**

43 A deeper depression: **e. Fossa**

44 A groove: **j. Sulcus**

45 A hole: **d. Foramen**

46 A canal or opening to a canal: **g. Meatus**

47 A smaller, rounded eminence: **l. Tubercle**

48–56 Following is how Figure 5-1, the long bone, should be labeled.

48. **h. Articular cartilage;** 49. **d. Spongy bone tissue;** 50. **g. Red bone marrow;** 51. **i. Compact bone tissue;** 52. **e. Medullary or nutrient artery;** 53. **b. Medullary cavity;** 54. **f. Proximal epiphysis;** 55. **a. Diaphysis;** 56. **c. Distal epiphysis**

57–71 Following is how Figure 5-2, the lateral view of the skull, should be labeled.

57. **l. Coronal suture;** 58. **h. Parietal bone;** 59. **f. Squamosal suture;** 60. **a. Temporal bone;** 61. **i. Lambdoidal suture;** 62. **j. Occipital bone;** 63. **m. Mandibular condyle;** 64. **d. Styloid process;** 65. **o. Mandible;** 66. **g. Maxilla;** 67. **k. Zygomatic bone;** 68. **n. Zygomatic process;** 69. **c. Nasal bone;** 70. **b. Sphenoid bone;** 71. **e. Frontal bone**

72 Frontal and parietals: **d. Coronal suture**

73 Occipital and parietals: **b. Lambdoidal suture**

74 Parietal and parietal: **c. Sagittal suture**

75 Temporal and parietal: **a. Squamosal suture**

76–**82** Following is how Figure 5-3, the inferior view of the skull, should be labeled.

76. **f. Maxilla;** 77. **d. Palatine bone;** 78. **a. Vomer;** 79. **g. Sphenoid bone;** 80. **b. Mandibular fossa;** 81. **e. Occipital condyle;** 82. **c. Foramen magnum**

83–**88** Following is how Figure 5-4, the frontal view of the skull, should be labeled.

83. **d. Sphenoid bone;** 84. **f. Maxilla;** 85. **a. Zygomatic bone;** 86. **e. Lacrimal bone;** 87. **c. Ethmoid bone;** 88. **b. Vomer**

89–**105** Following is how Figure 5-5, the cranial cavity, should be labeled.

89. **k. Olfactory foramina;** 90. **p. Optic foramen (canal);** 91. **l. Foramen rotundum;** 92. **c. Foramen ovale;** 93. **e. Foramen lacerum;** 94. **a. Internal auditory (acoustic) meatus;** 95. **q. Jugular foramen;** 96. **m. Foramen magnum;** 97. **n. Crista galli;** 98. **f. Cribriform plate;** 99. **d. Frontal bone;** 100. **o. Sphenoid bone;** 101. **g. Sella turcica;** 102. **h. Temporal bone;** 103. **i. Foramen spinosum;** 104. **b. Parietal bone;** 105. **j. Occipital bone**

106–**109** Following is how Figure 5-6, the sinus view of the skull, should be labeled.

106. **b. Frontal sinus;** 107. **d. Ethmoid sinus;** 108. **a. Sphenoid sinus;** 109. **c. Maxillary sinus**

110–**118** Following is how Figure 5-7, the vertebral column, should be labeled.

110. **f. Cervical vertebrae or curvature;** 111. **c. Thoracic vertebrae or curvature;** 112. **g. Lumbar vertebrae or curvature;** 113. **d. Sacrum;** 114. **a. Coccyx;** 115. **b. Intervertebral foramen;** 116. **i. Intervertebral disc;** 117. **e. A vertebra;** 118. **h. A spinous process**

119–**127** Following is how Figure 5-8, the vertebra, should be labeled.

119. **a. Vertebral foramen;** 120. **i. Pedicle;** 121. **c. Transverse process;** 122. **g. Transverse foramen;** 123. **d. Spinous process (bifid);** 124. **b. Lamina;** 125. **h. Inferior articulating facet;** 126. **e. Superior articulating facet;** 127. **f. Body**

128–**133** Following is how Figure 5-9, the thoracic cage, should be labeled.

128. **f. Manubrium;** 129. **c. Body;** 130. **a. Xiphoid process;** 131. **b. Jugular notch;** 132. **e. Clavicular notch;** 133. **d. Costal cartilage**

134–**139** Fill in the blanks to complete the following sentences:

The organs protected by the thoracic cage include the **134. heart** and the **135. lungs.** The first seven pairs of ribs attach to the sternum by the costal cartilage and are called **136. true** ribs. Pairs 8 through 10 attach to the costal cartilage of the seventh pair and not directly to the sternum, so they're called **137. false** ribs. The last two pairs, 11 and 12, are unattached anteriorly, so they're called **138. floating** ribs. There's one bone in the entire skeleton that doesn't articulate with any other bones but nonetheless is considered part of the axial skeleton. It's called the **139. hyoid** bone.

140–**159** Following is how Figure 5-10, the frontal view of the skeleton, should be labeled.

140. **h. Clavicle;** 141. **c. Scapula;** 142. **k. Humerus;** 143. **g. Ilium;** 144. **l. Pubis;** 145. **t. Ishium;** 146. **n. Os coxa;** 147. **o. Sacrum;** 148. **s. Femur;** 149. **j. Patella;** 150. **d. Metatarsals;** 151. **p. Phalanges of the hand;** 152. **q. Metacarpals;** 153. **e. Carpals;** 154. **m. Radius;** 155. **b. Ulna;** 156. **a. Tibia;** 157. **i. Fibula;** 158. **r. Tarsals;** 159. **f. Phalanges of the foot**

160 Where do the clavicles articulate with the scapulae? **a. At the acromion process.**

161 Where does the biceps muscle attach on the radius? **c. The radial tuberosity.**

162 What structure of the humerus articulates with the head of the radius? **d. The capitulum.**

163   The patella is what kind of bone? **b. Sesamoid.**

164   Which of these bones is not part of the pelvic girdle? **b. Lumbar vertebrae.**

165   What is the formal term for the prominence commonly referred to as the elbow?
a. Olecranon process.

166   Where does the ulna articulate with the humerus? **d. The trochlea.**

167   What is the formal term for the socket at the head of the femur? **b. Acetabulum.**

168   Which is the largest and strongest tarsal bone? **d. The calcaneus.**

169   Where are people who complain about their sacroiliac experiencing pain? **a. Lower back.**

170   Sutures of the skull: **a. Fibrous joint.**

171   Fluid-filled cavity: **d. Synovial joint.**

172   Knee joint: **d. Synovial joint.**

173   Symphysis pubis: **c. Cartilaginous joint-symphysis.**

174   Epiphyseal plate: **b. Cartilaginous joint-synchondrosis.**

175   Intervertebral discs: **c. Cartilaginous joint-symphysis.**

176–180   Following is how Figure 5-11, the synovial joint, should be labeled.

176. **b. Periosteum;** 177. **e. Fibrous capsule;** 178. **a. Synovial (joint) cavity;** 179. **d. Articular cartilage;** 180. **c. Synovial membrane**

181   What is the formal term for an immovable joint? **b. A synarthrosis.**

182   What is the formal term for a freely moving joint? **c. A diarthrosis.**

183   What material or structure allows free movement in a joint? **c. Synovial fluid.**

184   Choose an example of a ball-and-socket joint. **b. The hip.**

185   Choose an example of a pivotal joint. **a. Between the radius and the ulna.**

186   Choose an example of a saddle joint. **b. The carpometacarpal joint of the thumb.**

187   Choose an example of a shoulder joint ligament. **a. The coracohumeral ligament.**

188   Choose an example of a knee joint ligament. **c. Oblique popliteal ligament.**

189   Choose an example of a hip joint ligament. **b. Pubofermoral ligament.**

190   What is the term for the structure in the knee that divides the synovial joint into two separate compartments? **d. Meniscus or articular disc.**

191   Flexion: **b. Decrease in the angle between two bones.**

192   Extension: **e. Increase in the angle between two bones.**

193   Abduction: **g. Movement away from the midline of the body.**

194   Adduction: **j. Movement toward the midline of the body.**

195   Rotation: **i. Turning around an axis.**

196   Pronation: **d. Downward or palm downward.**

197   Supination: **a. Upward or palm upward.**

198   Eversion: **f. Turning of the sole of the foot outward.**

199   Inversion: **c. Turning of the sole of the foot inward.**

200   Circumduction: **h. Forming of a cone with the arm or leg.**

Chapter 6

Getting in Gear: The Muscles

∙∙∙

In This Chapter
▶ Understanding the functions and structure of muscles

▶ Classifying types of muscle

▶ Getting the scoop on muscles as organs

▶ Breaking down muscle contractions, tone, and power

▶ Deciphering muscle names

∙∙∙

Much of what we think of as "the body" centers around our muscles and what they can do, what we want them to do, and how tired we get trying to make them do it. With all that muscles do and are, it's hard to believe the word "muscle" is rooted in the Latin word *musculus,* which is a diminutive of the word for "mouse." Well, the muscle is a mouse that roars. Muscles make up most of the fleshy parts of the body and average 43 percent of the body's weight. Layered over the skeleton (which we discuss in Chapter 5), they largely determine the body's form. More than 500 muscles are large enough to be seen by the unaided eye, and thousands more are visible only through a microscope. Although there are three distinct types of muscle tissue, every muscle in the human body shares one important characteristic: contractility, the ability to shorten, or contract.

Flexing Your Muscle Knowledge

The study of muscles is called *myology* after the Greek word *mys,* which means "mouse." Muscles perform a number of functions vital to maintaining life, including

✔ **Movement:** Skeletal muscles (those attached to bones) convert chemical energy into mechanical work, producing movement ranging from finger tapping to a swift kick of a ball by contracting, or shortening. Reflex muscle reactions protect your fingers when you put them too close to a fire and startle you into watchfulness when an unexpected noise sounds. Many purposeful movements require several sets, or groups, of muscles to work in unison.

✔ **Vital functions:** Without muscle activity, you die. Muscles are doing their job when your heart beats, when your blood vessels constrict, and when your intestines squeeze food along your digestive tract in *peristalsis.*

✔ **Antigravity:** Perhaps that's overstating it, but muscles do make it possible for you to stand and move about in spite of gravity's ceaseless pull. Did your mother tell you to improve your posture? Just think how bad it would be without any muscles!

- **Heat generation:** You shiver when you're cold and stamp your feet and jog in place when you need to warm up. That's because chemical reactions in muscles result in heat, helping to maintain the body's temperature.

- **Keeping the body together:** Muscles are the warp and woof of your body's structure, binding one part to another.

- **Joint stability:** Muscles and their tendons aid the ligaments to reinforce joints.

- **Supporting and protecting soft tissues:** The body wall and floor of the pelvic cavity support the internal organs and protect soft tissues from injury.

- **Sphincters:** These muscles encircle openings to control swallowing, defecation, and urination.

As you may remember from studying tissues (see Chapter 4 for details), muscle cells — called *fibers* — are some of the longest in the body. Fibers are held together by connective tissue and enclosed in a fibrous sheath called *fascia.* Some muscle fibers contract rapidly, whereas others move at a leisurely pace. Generally speaking, however, the smaller the structure to be moved, the faster the muscle action. Exercise can increase the thickness of muscle fibers, but it doesn't make new fibers. Skeletal muscles have a rich vascular supply that dilates during exercise to give the working muscle the extra oxygen it needs to keep going.

Two processes are central to muscle development in the developing embryo: *myogenesis,* during which muscle tissue is formed, and *morphogenesis,* when the muscles form into internal organs. By the eighth week of gestation, a fetus is capable of coordinated movement.

Following are some important muscle terms to know:

- **Fascia:** Loose, or areolar, connective tissue that holds muscle fibers together to form a muscle organ

- **Fiber:** An individual muscle cell

- **Insertion:** The more movable attachment of a muscle

- **Ligament:** Dense, regular connective tissue that supports joints and anchors organs

- **Motor nerve:** A nerve that stimulates contraction of a muscle

- **Myofibril:** Fibrils within a muscle cell that contain protein filaments such as actin and myosin that slide during contraction, shortening the fiber (or cell)

- **Origin:** The immovable attachment of a muscle, or the point at which a muscle is anchored by a tendon to the bone

- **Sarcoplasm:** The cellular cytoplasm in a muscle fiber

- **Tendon:** Connective tissue made up of collagen, a fibrous protein that attaches muscles to bone; allows some muscles to apply their force at some distance from where a contraction actually takes place

- **Tone, or tonus:** State of tension present to a degree at all times, even when the muscle is at rest

Complete the following practice questions to see how well you understand the basics of myology:

1. Which of the following is *not* a true statement?

   a. Muscle fibers are some of the longest cells in the body.

   b. Myofibrils within muscle cells contain protein filaments that slide during contraction.

   c. Sphincters hold fibers together to form muscles.

   d. Posture is an expression of muscle action.

2. What happens during myogenesis and morphogenesis?

   a. Muscles that aren't flexed regularly begin to lose tone during myogenesis and are eaten away entirely during morphogenesis.

   b. Collagen is converted into muscle tissue during myogenesis and recycled by the body during morphogenesis.

   c. Chemical energy is converted into mechanical work during myogenesis and morphogenesis.

   d. Embryonic muscles form during myogenesis, and muscles form into internal organs during morphogenesis.

3. Why do your muscles shiver when you're cold?

   a. To prevent paralysis

   b. To generate heat

   c. To improve elasticity

   d. To maintain contractility

4. Name the cellular unit in muscle tissue.

   a. Filament

   b. Myofibril

   c. Fiber

   d. Fasciculus

5. What defines a state of tension present to a degree at all times?

   a. Rigor

   b. Tonus

   c. Clovus

   d. Paralysis

6. It's possible to completely relax every muscle in the body.

   a. True

   b. False

7. Exercise forms new muscle fibers.

   a. True

   b. False

Muscle Classifications: Smooth, Cardiac, and Skeletal

Muscle tissue is classified in three ways based on the tissue's function, shape, and structure:

✓ **Smooth muscle tissue:** So-called because it doesn't have the cross-striations typical of other kinds of muscle, the spindle-shaped fibers of smooth muscle tissue do have faint longitudinal striping. This muscle tissue forms into sheets and makes up the walls of hollow organs such as the stomach, intestines, and bladder. The tissue's involuntary movements are relatively slow, so contractions last longer than those of other muscle tissue, and fatigue is rare. Each fiber is about 6 microns in diameter and can vary from 15 microns to 500 microns long. If arranged in a circle inside an organ, contraction constricts the cavity inside the organ. If arranged lengthwise, contraction of smooth muscle tissue shortens the organ.

✓ **Cardiac muscle tissue:** Found only in the heart, cardiac muscle fibers are cylindrically branched, cross-striated, feature one central nucleus, and move through involuntary control. An electron microscope view of the tissue shows separate fibers tightly pressed against each other, forming cellular junctions called *intercalated discs* that look like tiny, dark-colored plates. It's believed that intercalated discs are not just cellular junctions but special structures that help move an electrical impulse throughout the heart.

✓ **Skeletal muscle tissue:** This is the tissue that most people think of as muscle. It's the only muscle subject to control by the central nervous system via the somatic nervous system (SoNS). The somatic nervous system (or voluntary nervous system) is the part of the peripheral nervous system that provides the motor innervation needed for voluntary control of body movement via skeletal muscles (see Chapter 15 for an introduction to the nervous system).

The long, striated cylindrical fibers of skeletal muscle tissue contract quickly but tire just as fast. Skeletal muscle, which is also what's considered meat in animals, is 20 percent protein, 75 percent water, and 5 percent organic and inorganic materials. Each multinucleated fiber is encased in a thin, transparent cell membrane called a *sarcolemma* that receives and conducts stimuli. The fibers, which vary from 10 microns to 100 microns in diameter and up to 4 centimeters in length, are subdivided lengthwise into tiny myofibrils roughly 1 micron in diameter that are suspended in the cell's sarcoplasm.

The following practice questions test your knowledge of muscle classifications:

8. What type of muscle tissue lacks cross-striations?

   a. Cardiac

   b. Smooth

   c. Skeletal

   d. Contracting

9. What does a sarcolemma do?

   a. It receives and conducts stimuli for skeletal muscle tissue.

   b. It provides a structure for moving electrical impulses through the heart.

   c. It forms a wall for hollow organs.

   d. It prevents muscle fatigue.

10. Which muscle type appears only in a single organ?

 a. Contractile

 b. Smooth

 c. Cardiac

 d. Skeletal

11. Intercalated discs

 a. anchor skeletal muscle fibers to one another.

 b. may play a role in moving electrical impulses through the heart.

 c. are found only in the muscles of the back.

 d. contribute to tactile perception.

Contracting for a Contraction

Before we can explain how and why muscles do what they do, it's important that you understand the anatomy of how they're put together. Use Figure 6-1 as a visual guide as you read through the following sections.

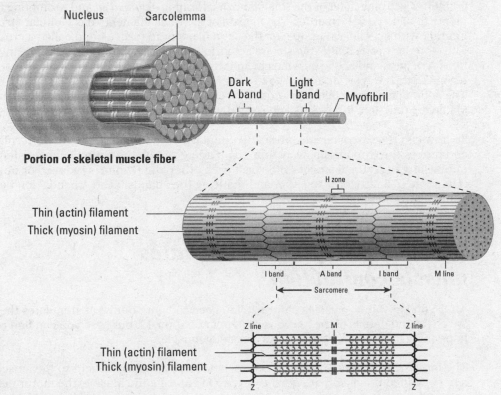

Figure 6-1: Microscopic anatomy of a skeletal muscle fiber.

Illustration by Imagineering Media Services Inc.

Breaking down a muscle's anatomy and movement

We base this description of muscle on the most studied classification of muscle: skeletal. Each fiber packed inside the sarcolemma contains hundreds, or even thousands, of myofibril strands made up of alternating filaments of the proteins *actin* and *myosin*. Actin and myosin are what give skeletal muscles their striated appearance, with alternating dark and light bands. The dark bands are called *anisotropic,* or *A bands*. The light bands are called *isotropic,* or *I bands*. In the center of each I band is a line called the *Z line* that divides the myofibril into smaller units called *sarcomeres*. At the center of the A band is a less-dense region called the *H zone*. The H zone contains the *M line,* a fine filamentous structure that holds the thick myosin filaments in parallel arrangement.

Now, this is where the actin and myosin come in. Each sarcomere contains thick filaments of myosin in the A band and thin filaments of actin primarily in the I band but extending a short distance between the myosin filaments into the A band. In a noncontracting fiber, actin filaments do not extend into the central area of the A band. This explains why the H zone is less dense. Those thin actin filaments are anchored to the Z line at their midpoints, which holds them in place and creates a structure against which the filaments exert their pull during contraction.

The theory of contraction called the *Interdigitating Filament Model of Muscle Contraction,* or the *Sliding Filament Theory of Muscle Contraction,* says that the myosin of the thick filaments combines with the actin of the thin filaments, forming *actomyosin* and prompting the filaments to slide past each other. The myosin of the thick filaments has globular structures that interact with special active sites on the actin filament to form a bond called a crossbridge. The globular heads of the myosin are active in moving the actin filaments. The myosin contains ATP sites and active enzymes to split the ATP and release energy for the reaction. (If the term *ATP* throws you for a loop, go to Chapter 1 for a look at the basic chemistry of biology). The actin filament contains the protein tropomysin that plays a critical role in regulating the filament's function. It prevents and regulates interaction of actin and myosin.

So how can the sarcomeres contract? Troponin, a second protein, binds with calcium ions and moves the tropomysin away from the binding site on the actin filament, effectively unblocking it. As the filaments slide past each other, the H zone is reduced or obliterated, pulling the Z lines closer together and reducing the I bands. (The A bands don't change.) Voilà! Contraction has occurred!

Understanding what stimulates muscle contraction

After you know how muscles contract, you need to figure out what stimulates them to do so. We cover the details of the nervous system in Chapter 15, but here you can find out what's happening as an impulse stimulates a skeletal muscle.

The *impulse,* or *stimulus,* from the central nervous system is brought to the muscle through a nerve called the *motor,* or *efferent,* nerve. On entering the muscle, the motor nerve fibers separate to distribute themselves among the thousands of muscle fibers. Because the muscle has more fibers than the motor nerve, individual nerve fibers branch repeatedly so that a single nerve fiber innervates from 5 to as many as 200 muscle fibers. These small terminal branches penetrate the sarcolemma and form a special structure known as the *motor end plate,* or *synapse.* This neuromuscular unit consisting of one motor neuron and all the muscle fibers that it innervates is called the *motor unit.*

Interference — either chemical or physical — with the nerve pathway can affect the action of the muscle or stop the action altogether, resulting in muscle paralysis. There also are *afferent,* or *sensory,* nerves that carry information about muscle condition to the brain.

When an impulse moves through the synapse and the motor unit, it must arrive virtually simultaneously at each of the individual sarcomeres to create an efficient contraction. Enter the *transverse system,* or *T system,* of tubules. The fiber's membrane forms deep *invaginations,* or inward-folding sheaths, at the Z line of the myofibrils. The resulting inward-reaching tubules ensure that the sarcomeres are stimulated at nearly the same time.

Does it matter whether the signal received is strong or weak? Nope. That's the *all-or-none law* of muscle contraction. The fiber either contracts completely or not at all. In other words, if a single muscle fiber is going to contract, it's going to do so to its fullest extent.

Following are some practice questions that deal with muscle anatomy and contraction:

12.–16. Match each muscle component with the appropriate region.

12. _____ Myosin   **a.** H zone

13. _____ Segment of fibril from Z line to Z line   **b.** Z line

14. _____ Less-dense region of the A band   **c.** I band

15. _____ Structure to which filaments are attached   **d.** A band

16. _____ Actin   **e.** Sarcomere

17. Which of these terms doesn't belong in the following list?

a. Anisotropic

b. Actin

c. Myosin

d. Sarcolemma

18. During contraction, what does *not* occur?

a. Muscle fibers stretch.

b. Thick and thin filaments slide past each other.

c. Muscle fibers shorten.

d. Troponin binds with calcium ions.

19. A weak stimulus causes a muscle fiber to contract only partway.

a. True

b. False

Pulling Together: Muscles as Organs

A muscle organ has two parts:

- ✔ **The belly:** Composed predominantly of muscle fibers

- ✔ **The tendon:** Composed of fibrous, or collagenous, regular connective tissue. If the tendon is a flat, sheetlike structure attaching a wide muscle, it's called an *aponeurosis*.

The sarcolemma of each muscle fiber is surrounded by areolar connective tissue called *endomysium* that binds the fibers together into bundles called *fasciculi* (see Figure 6-2). Each bundle, or *fasciculus,* is surrounded by areolar connective tissue called *perimysium.* All the fasciculi together make up the belly of the muscle, which is surrounded by dense connective tissue called the *epimysium.* Blood vessels, lymph vessels, and nerves pass into the fasciculus through areolar connective tissue called the *trabecula.* These blood vessels in turn branch off into capillaries that surround the muscle fibers in the endomysium.

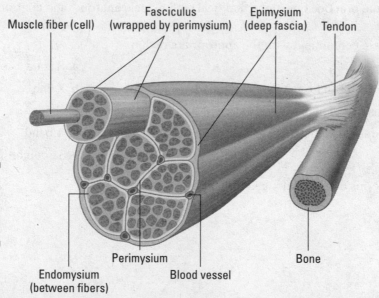

Figure 6-2:
Connective
tissue in a
muscle.

Muscle fiber (cell) — Fasciculus (wrapped by perimysium) — Epimysium (deep fascia) — Tendon

Perimysium — Bone

Endomysium (between fibers) — Blood vessel

Illustration by Imagineering Media Services Inc.

20.–24. Match the muscle structures with their descriptions.

20. _____ Areolar connective tissue that bundles fibers

21. _____ Bundles of muscle fibers

22. _____ Connective tissue that surrounds a bundle of muscle fibers

23. _____ Connective tissue through which arteries and veins enter muscle bundles

24. _____ Flat, sheetlike tendon that serves as insertion for a large, flat muscle

a. Perimysium

b. Aponeurosis

c. Trabecula

d. Fasciculi

e. Endomysium

Assuming the Right Tone

As we note in the earlier section "Understanding what stimulates muscle contraction," when it comes to contraction of a muscle fiber, it's an all-or-nothing affair. Nonetheless, the stimuli may vary in frequency and spatial distribution. It has been demonstrated that an increase in the frequency of stimulatory nerve impulses gives a stronger contraction. Maximum stimulus brings all motor units to bear together. Conversely, slower or fewer action potentials — a weaker stimulus, as it were — causes fewer motor units to become involved in a contraction. So it's true that a muscle organ can have graded degrees of contraction depending on the level of stimulation. As for how this can be so, one theory proposes that individual fibers have specific thresholds of excitation; thus, those with higher thresholds respond only to stronger stimuli. Another theory holds that the deeper a fiber is buried in the muscle, the less accessible it is to incoming stimuli.

In physiology, a muscle contraction is referred to as a *muscle twitch*. A twitch is the fundamental unit of recordable muscular activity. The twitch consists of a single stimulus-contraction-relaxation sequence in a muscle fiber. The contraction is the period when the crossbridges are active. The relaxation phase occurs when crossbridges detach and the muscle tension decreases. Complete fatigue occurs when no more twitches can be elicited, even with increasing intensity of stimulation.

The short lapse of time between the application of a stimulus and the beginning of muscular response is called the *latent period* or *phase*. In mammalian muscle, latency is about .001 second, or one one-thousandth of a second.

Several types of muscle contractions relate to tone:

- **Isometric:** Occurs when a contracting muscle is unable to move a load (or heft a piece of luggage or push a building to one side). It retains its original length but develops *tension*. No mechanical work is accomplished, and all energy involved is expended as heat.

- **Isotonic:** Occurs when the resistance offered by the load (or the gardening hoe or the cold can of soda) is less than the tension developed, thus shortening the muscle and resulting in mechanical work. There are two main types of isotonic contraction:

  - **Concentric contraction:** Occurs when the force generated by the muscle is less than the maximum tension. The muscle shortens with increased velocity. An example is weight lifting that performs work on a load.

  - **Eccentric contraction:** Occurs when the force of a load is greater than the maximum muscle tension, causing the muscle to elongate and performing work on the muscle. Muscle injury and soreness may result.

But muscles aren't independent sole proprietors. Each muscle depends upon companions in a muscle group to assist in executing a particular movement. That's why muscles are categorized by their actions. The brain coordinates the following groups through the cerebellum.

- **Prime movers:** Provide the major force for producing a specific movement. Just as it sounds, these muscles are the workhorses that produce movement.

- **Antagonists:** These muscles exist in opposition to prime movers, regulating the motion by contracting and providing resistance.

> ✔ **Fixators or fixation muscles:** These muscles serve to steady a part while other muscles execute movement. They don't actually take part in the movement itself.

> ✔ **Synergists:** These muscles control movement of the proximal joints so that the prime movers can bring about movements of distal joints.

Flex your knowledge of muscle tone and function with these practice questions:

Q. Muscle contraction that is unable to move an object involves an action known as

   a. isometric.

   b. eccentric.

   c. isotonic.

   d. concentric.

A. The correct answer is isometric. When the tension leads to movement (actual work), it's isotonic. Eccentric contraction and concentric contraction are the two types of isotonic contraction.

25. What do you call muscles that tend to counteract or slow an action?

   a. Antagonists

   b. Fixators

   c. Primary movers

   d. Synergists

26. What creates a strong muscle contraction?

   a. Isotonic influences building up as mechanical work

   b. Isometric influences building up as tension

   c. Low frequency stimulatory nerve impulses

   d. High frequency stimulatory nerve impulses

27. What is a muscle contraction called?

   a. Latency

   b. Synergy

   c. A twitch

   d. Isotonic motion

Leveraging Muscular Power

Skeletal muscle power is nothing without lever action. The bone acts as a rigid bar, the joint is the fulcrum, and the muscle applies the force. Levers are divided into the *weight arm,* the area between the fulcrum and the weight; and the *power arm,* the area between the fulcrum and the force. When the power arm is longer than the weight arm, less force is required to lift the weight, but range, or distance, and speed are sacrificed. When the weight arm is longer than the power arm, the range of action and speed increase, but power is sacrificed. Therefore, 90 degrees is the optimum angle for a muscle to attach to a bone and apply the greatest force.

Three classes of levers are at work in the body:

- **Class I, or seesaw:** The fulcrum is located between the weight and the force being applied. An example is a nod of the head: The head-neck joint is the fulcrum, the head is the weight, and the muscles in the back of the neck apply the force.

- **Class II, or wheelbarrow:** The weight is located between the fulcrum and the point at which the force is applied. An example is standing on your tiptoes: The fulcrum is the joint between the toes and the foot, the weight is the body, and the muscles in the back of the leg at the heel bone apply the force.

- **Class III, or removing a nail with a hammer:** The force is located between the weight and the fulcrum. An example is flexing your arm and showing off your biceps: The elbow joint is the fulcrum, the weight is the lower arm and hand, and the biceps insertion on the lower arm applies the force.

The direction in which the muscle fibers run also plays a critical role in leverage. Here are the possible directions:

- **Longitudinal:** Fibers run parallel to each other, or longitudinally, the length of the muscle. Example: sartorius.

- **Pennate:** Fibers attach to the sides of the tendon, extending the length of the muscle. These come in subcategories:

  - **Unipennate:** Fibers attach to one side of the tendon; example: tibialis posterior
  - **Bipennate:** Fibers attach to two sides of the tendon; example: rectus femoris
  - **Multipennate:** Fibers attach to many sides of the tendon; example: deltoid

- **Radiate:** Fibers converge from a broad area into a common point. Example: pectoralis major.

- **Sphincter:** Fibers are arranged in a circle around an opening. Example: orbicularis oculi.

The three types of *fasciae,* which *Gray's Anatomy* describes as "dissectable, fibrous connective tissues of the body," are as follows:

- **Superficial fasciae:** Found under the skin and consisting of two layers: an outer layer called the *panniculus adiposus* containing fat; and an inner layer made up of a thin, membranous, and highly elastic layer. Between the two layers are the superficial arteries, veins, nerves, and mammary glands.

- **Deep fasciae:** Holds muscles or structures together or separates them into groups that function in unison. It's a system of splitting, rejoining, and fusing membranes involving

  - An outer investing layer that's found under the superficial fasciae covering a large part of the body
  - An internal investing layer that lines the inside of the body wall in the torso, or trunk, region
  - An intermediate investing layer that connects the outer investing layer and the internal investing layer

- **Subserous fasciae:** Located between the internal investing layer of the deep fasciae and the peritoneum. It's the subserous membrane that lines the *abdominopelvic* cavity, also known as the *peritoneal cavity.*

Got all that? Then try your hand at the following questions:

28. Which of the following in Figure 6-3 is a Class II lever?

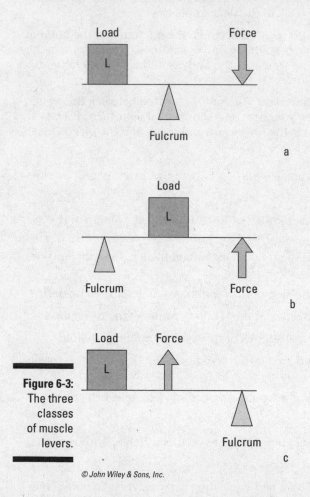

Figure 6-3:
The three
classes
of muscle
levers.

© John Wiley & Sons, Inc.

29. Which muscles would provide the force in a Class III lever?

a. Biceps brachii

b. Pectoralis major

c. Orbicularis oculi

d. Sartorius

30. Which of the following would produce a wide range of movement with speed while sacrificing power?

a. Power arm and weight arm of equal lengths

b. Long weight arm, short power arm

c. Long power arm, shorter weight arm

d. Short power arm, shorter weight arm

31. Which of the following is a bipennate bundle arrangement?

   a. Sartorius

   b. Rectus femoris

   c. Pectoralis major

   d. Tibialis posterior

32. Which of the following is considered a dissectable connective tissue?

   a. Aponeurosis

   b. Bursae

   c. Fasciae

   d. Tendons

33. What do deep fasciae do?

   a. Contain vascular structures within the panniculus adiposus

   b. Hold muscles or structures together, or separate them into groups that function in unison

   c. Line the peritoneal cavity

   d. Provide a structure for the abdominopelvic cavity

What's in a Name? Identifying Muscles

It may seem like a jumble of meaningless Latin at first, but muscle names follow a strict convention that names them according to one or more of the following:

   ✔ **Function:** These muscle names usually have a verb root and end in a suffix (*–or* or *–eus*), followed by the name of the affected structure. Example: levator scapulae (elevates the scapulae).

   ✔ **Compounding points of attachment:** These muscle names blend the origin and insertion attachment with an adjective suffix (*–eus* or *–is*). Examples: sternocleidomastoideus (sternum, clavicle, and mastoid process) and sternohyoideus (sternum and hyoid).

   ✔ **Shape or position:** These muscle names usually have descriptive adjectives that may be followed by the names of the locations of the muscles. Examples: rectus (straight) femoris, rectus abdominis, and serratus (sawtooth) anterior.

   ✔ **Figurative resemblance:** These muscle names are based on the muscles' resemblance to some objects. Examples: gastrocnemius (resembles the stomach) and trapezius (resembles a trapezoid or kite shape).

Check out Figure 6-4 and Table 6-1 for a rundown of prominent muscles in the body and key points to remember about each one.

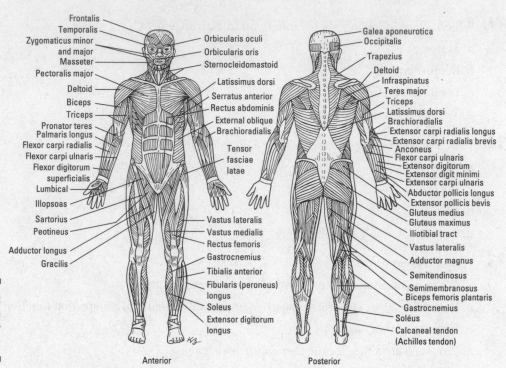

Figure 6-4:
Some of the body's major muscles.

Anterior   Posterior

Illustration by Kathryn Born, MA

Table 6-1		Muscles of the Body	
Muscle	**Origin**	**Insertion**	**Action**
Head			
Frontalis	Galea aponeurotica	Eyebrow	Expression
Occipitalis	Occipital bone	Galea aponeurotica	Moves scalp forward and backward
Buccinator	Alveolar processes of mandible and maxillary bone	Orbicularis oris and skin at angle of mouth	Mastication
Orbicularis oculi	Encircles eye	Encircles orbit and extends within eyelid	Closes eye
Orbicularis oris	Encircles mouth	Skin surrounding mouth and at angle of mouth	Closes mouth
Masseter	Zygomatic arch	Mandible	Mastication
Temporalis	Temporal fossa	Coronoid process and ramus of mandible	Mastication
Zygomaticus	Zygomatic bone	Corner of mouth	Smiling

Muscle	Origin	Insertion	Action
Neck			
Sternocleidomastoid	Manubrium of sternum and median portion of clavicle	Mastoid process of temporal bone	Rotation and flexion of neck vertebrae
Back			
Latissimus dorsi	Vertebral column	Humerus	Extends at shoulder joint
Trapezius	Occipital bone and vertebral column (7 cervical vertebrae and all thoracic vertebrae)	Clavicle and scapula	Rotates, elevates, depresses, adducts, and stabilizes scapula
Pectoral girdle			
Pectoralis major	Sternum, clavicle, upper 6 ribs, and aponeurosis of external oblique muscle	Medial margin of tertubercular groove of humerus	Flexion, rotation, and adduction of shoulder joint
Shoulder			
Deltoid	Lateral third of clavicle, acromion process, and spine of scapula	Deltoid tuberosity of humerus	Abduction, flexion, rotation, and extension of arm, using coordinated fibers: entire muscle (abduction); interior fibers (flexion); medial fibers (rotation); posterior fibers (extension); and lateral fibers (rotation) of arm
Abdominal wall			
External abdominal oblique	External surface of lower 8 ribs	Aponeurosis to linea alba and anterior half of iliac crest	Stabilizes, protects, and supports internal viscera, compresses abdominal cavity, assists in flexing and rotating vertebral column
Internal abdominal oblique	Inguinal ligament, iliac crest, and lumbodorsal fasciae	Linea alba, pubic crest, and last 3 ribs	Stabilizes, protects, and supports internal viscera, compresses abdominal cavity, assists in flexing and rotating vertebral column

(continued)

Table 6-1 *(continued)*

Muscle	Origin	Insertion	Action
Transversus abdominis	Inguinal ligament, iliac crest, lumbodorsal fasciae, and costal cartilage of last 6 ribs	Linea alba and pubic crest	Stabilizes, protects, and supports internal viscera, compresses abdominal cavity, assists in flexing and rotating vertebral column
Rectus abdominis	Pubic crest and symphysis pubis	Xiphoid process of sternum and costal cartilage of ribs 5–7	Stabilizes, protects, and supports internal viscera
Thorax			
Diaphragm	Xiphoid process of sternum, inner surface of lower 6 ribs, and lumbar vertebrae	Central tendon of diaphragm	Pulls central tendon downward, increasing size of thoracic cavity and causing inspiration; separates thoracic and abdominal cavities
External intercostals	Inferior border of rib above and costal cartilage	Superior border of rib below rib of origin	Elevates ribs, aiding inspiration
Internal intercostals	Superior border of rib below and costal cartilage	Inferior border of rib above rib of origin	Depresses ribs, aiding inspiration
Arm			
Biceps brachii	Long head: Tubercle above glenoid fossa Short head: Coracoid process of scapula	Tuberosity of radius	Flexion at elbow joint
Triceps brachii	Long head: Infraglenoid tubercle of scapula Lateral head: Posterior surface of humerus above radial groove Medial head: Posterior surface of humerus above radial groove	Olecranon process of ulna	Extension of elbow joint

Muscle	Origin	Insertion	Action
Flexor carpi radialis	Medial epicondyle of humerus	2nd to 3rd metacarpals	Flexor of wrist, abducts hand
Flexor carpi ulnaris	Medial epicondyle of humerus and olecranon process	Pisiform bone and base of 5th metacarpal	Flexor of wrist, adducts hand
Supinator	Lateral epicondyle of humerus and proximal end of ulna	Proximal end of radius	Supinates forearm
Extensor carpi ulnaris	Lateral epicondyle of humerus	Base of 5th metacarpal	Extends and adducts wrist
Extensor carpi radialis longus	Lateral supracondylar ridge of humerus	Base of 2nd metacarpal	Extends and abducts wrist
Extensor carpi radialis brevis	Lateral epicondyle of humerus	Dorsal surface of 3rd metacarpal	Extends and abducts wrist, steadies wrist during finger flexion
Thigh, Anterior (Front)			
Rectus femoris (part of quadriceps)	Acetabulum of coxal bone	Tibial tuberosity via patella	Extends knee and flexes thigh at hip
Vastus lateralis (part of quadriceps), vastus medialis, vastus intermedialis (part of quadriceps)	Femur	Tibia	Extends knee
Sartorius	Anterior superior iliac spine	Proximal medial surface of tibia	Flexes at knee and laterally rotates thigh
Adductors	Pubis	Femur	Adduction and flexion at hip joint and lateral rotation of thigh
Gracilis	Pubis, symphysis pubis, and pubic arch	Medial surface of tibia	Adduction of thigh and rotation of leg
Thigh, Posterior (Back)			
Biceps femoris (part of hamstrings)	Long head: Ischial tuberosity Short head: Linea aspera	Fibula and lateral condyle of tibia	Flexion at knee, extension of thigh
Semimembranosus (part of hamstrings)	Ischial tuberosity	Medial condyle of tibia	Flexion at knee, extension of thigh
Semitendinosus (part of hamstrings)	Ischial tuberosity	Proximal end of tibia	Flexion at knee, extension of thigh

(continued)

Table 6-1 *(continued)*

Muscle	Origin	Insertion	Action
Leg, Posterior			
Gastrocnemius	Medial and lateral condyles of femur	Calcaneus by Achilles (calcaneal) tendon	Flexion at knee and plantar flexion
Soleus	Posterior third of fibula and middle third of tibia	Calcaneus by Achilles tendon	Plantar flexion
Hip			
Gluteus maximus	Dorsal ilium, sacrum, and coccyx	Glutual tuberosity of femur	Extends thigh and laterally rotates thigh

34. A muscle's name can be based on four things. Which one of those things was used to name these three muscles: the latissimus dorsi, the rectus abdominis, and the serratus anterior?

   a. Shape

   b. Attachment

   c. Figurative resemblance

   d. Function

35. The _____ of the sternocleidomastoid muscle gives it its name.

   a. function

   b. location

   c. attachment

   d. figurative resemblance

36. In humans, the origin of the biceps brachii would best include which of the following?

   a. Scapula

   b. Clavicle

   c. Fibula

   d. Ulna

37. Which of the following are insertions for the triceps and biceps brachii?

   a. Humerus and ulna

   b. Radius and humerus

   c. Scapula and humerus

   d. Radius and ulna

38.–42. Match the origins and insertions for the following muscles.

38. _____ Semimembranosus		**a.** The pubis and the femur
39. _____ Gracilis		**b.** The femur and the calcaneus
40. _____ Sartorius		**c.** The ilium and the tibia
41. _____ Gastrocnemius		**d.** The ischium and the tibia
42. _____ Adductors		**e.** The pubis and the tibia

43.–47. Match the muscles with their actions.

43. _____ Semitendinosus		**a.** Rotation of scapula
44. _____ Temporalis		**b.** Flexion of leg at knee joint
45. _____ Biceps brachii		**c.** Extension at shoulder joint
46. _____ Latissimus dorsi		**d.** Mastication
47. _____ Trapezius		**e.** Flexion of forearm

48.–52. Match the muscles with their locations.

48. _____ Latissimus dorsi		**a.** Head
49. _____ Internal oblique		**b.** Abdomen
50. _____ Quadriceps		**c.** Back
51. _____ Masseter		**d.** Neck
52. _____ Sternocleidomastoid		**e.** Thigh

53. Which of the following is *not* included in the quadriceps group?

a. Vastus intermedialis

b. Vastus lateralis

c. Rectus abdominis

d. Rectus femoris

54. What "two-headed" muscle is found in both the arm and the leg?

a. Quadriceps

b. Triceps

c. Biceps

d. Gluteus

55. This muscle divides the thoracic cavity from the abdominal cavity.

a. Diaphragm

b. External oblique

c. Transversus abdominis

d. Rectus abdominis

56. Together, the gastrocnemius and the soleus muscles are referred to as the triceps surae. Along with a third muscle, the plantaris, they contribute to the strongest tendon in the body, which is

a. the conjoint tendon.

b. the common flexor tendon.

c. the falx inguinalis.

d. the Achilles tendon.

57. Which of the following is *not* one of the muscles referred to as hamstrings?

a. Biceps femoris

b. Gracilis

c. Semimembranosus

d. Semitendinosus

Answers to Questions on Muscles

The following are answers to the practice questions presented in this chapter.

1 Which of the following is *not* a true statement? **c. Sphincters hold fibers together to form muscles.** Sphincters are muscular rings used as a control mechanism.

2 What happens during myogenesis and morphogenesis? **d. Embryonic muscles form during myogenesis, and muscles form into internal organs during morphogenesis.**

3 Why do your muscles shiver when you're cold? **b. To generate heat.** Chemical reactions in muscles result in heat, which is also why you stamp your feet and jog in place when you're cold.

4 Name the cellular unit in muscle tissue. **c. Fiber.** When it comes to muscle, fiber and cells are the same thing.

5 What defines a state of tension present to a degree at all times? **b. Tonus.** That's the elusive muscle "tone" for the flabby amongst us.

6 It's possible to completely relax every muscle in the body. **b. False.** If every muscle in the body were to relax, the heart would stop beating and food would stop moving through the digestive system.

7 Exercise forms new muscle fibers. **b. False.** Exercise can't form new fibers, but it can thicken what's already there.

8 What type of muscle tissue lacks cross-striations? **b. Smooth.** Without striations, this tissue can contract slowly for a very long time.

9 What does a sarcolemma do? **a. It receives and conducts stimuli for skeletal muscle tissue.** Other answers relate to other types of muscles, so if you associate "skeletal" with "sarcolemma," you'll stay on the right track.

10 Which muscle type appears only in a single organ? **c. Cardiac.** And that sole organ is the heart.

11 Intercalated discs **b. may play a role in moving electrical impulses through the heart.** There's evidence that these structures help keep the heart synchronized.

12 Myosin: **d. A band.**

13 Segment of fibril from Z line to Z line: **e. Sarcomere.**

14 Less-dense region of the A band: **a. H zone.**

15 Structure to which filaments are attached: **b. Z line.**

16 Actin: **c. I band.**

17 Which of these terms doesn't belong in the following list? **d. Sarcolemma.** This is the cell membrane encasing the myofibrils. All the other answer options refer to various protein structures.

18 During contraction, what does *not* occur? **a. Muscle fibers stretch.** Quite the opposite: Muscle fibers shorten.

19 A weak stimulus causes a muscle fiber to contract only partway. **b. False.** Muscle contractions are all-or-nothing; there's no such thing as a partial contraction.

20 Areolar connective tissue that bundles fibers: **e. Endomysium.**

21 Bundles of muscle fibers: **d. Fasciculi.**

22 Connective tissue that surrounds a bundle of muscle fibers: **a. Perimysium.**

23 Connective tissue through which arteries and veins enter muscle bundles: **c. Trabecula.**

24 Flat, sheetlike tendon that serves as insertion for a large, flat muscle: **b. Aponeurosis.**

25 What do you call muscles that tend to counteract or slow an action? **a. antagonists.**

TIP

The muscles are against the action, so think of them as antagonistic.

26 What creates a strong muscle contraction? **d. High frequency stimulatory nerve impulses.** The first two answers relate to types of muscle contractions, not strength.

27 What is a muscle contraction called? **c. A twitch.** Keep in mind the twitch is the fundamental measure of muscle activity.

28 Which of the following in Figure 6-3 is a Class II lever? **b.** With Class II, remember to look for wheelbarrow-like leverage.

29 Which muscles would provide the force in a Class III lever? **a. Biceps brachii.** It's the Popeye weight-lifting class, after all.

30 Which of the following would produce a wide range of movement with speed while sacrificing power? **b. Long weight arm, short power arm.** The longer the weight arm, the greater the range of action and speed *but* the less power there is.

31 Which of the following is a bipennate bundle arrangement? **b. Rectus femoris.** Keep in mind the muscle naming conventions we outline in the earlier section "What's in a Name? Identifying Muscles." Names are based on function, attachment, shape/position, or resemblance to a common object.

32 Which of the following is considered a dissectable connective tissue? **c. Fasciae.** This connective tissue can be found in most of the body.

33 What do deep fasciae do? **b. Hold muscles or structures together, or separate them into groups that function in unison.** All the other answers relate to the wrong types of fasciae.

34 A muscle's name can be based on four things. Which one of those things was used to name these three muscles: the latissimus dorsi, the rectus abdominis, and the serratus anterior? **a. Shape.** It's a matter of memorizing a bit of Latin: *latissimus* stems from the Latin word for "wide," *rectus* from the Latin word for "straight," and *serratus* from the Latin word for "notched" or "scalloped."

35 The **c. attachment** of the sternocleidomastoid muscle gives it its name. You can figure out this answer by recalling that *cleido* stems from the Latin word for "collarbone."

36 In humans, the origin of the biceps brachii would best include which of the following? **a. Scapula.** After you know that your biceps brachii are in your upper arm, you just have to remember the name of the origin bone. The Latin meaning of *scapulae* is, literally, "shoulder blades," so perhaps a better memory tool would be that the scapula looks vaguely like a spatula.

37 Which of the following are insertions for the triceps and biceps brachii? **d. Radius and ulna.** Once again, it's a matter of remembering where those muscles are — the upper arm — and then sorting out the names of the insertion points. It's helpful to remember that these two muscles are antagonists, working in opposition to one another.

38 Semimembranosus: **d. The ischium and the tibia.**

39 Gracilis: **e. The pubis and the tibia.**

40 Sartorius: **c. The ilium and the tibia.**

41 Gastrocnemius: **b. The femur and the calcaneus.**

42 Adductors: **a. The pubis and the femur.**

43 Semitendinosus: **b. Flexion of leg at knee joint.**

44 Temporalis: **d. Mastication**

45 Biceps brachii: **e. Flexion of forearm.**

46 Latissimus dorsi: **c. Extension at shoulder joint.**

47 Trapezius: **a. Rotation of scapula.**

48 Latissimus dorsi: **c. Back.**

49 Internal oblique: **b. Abdomen.**

50 Quadriceps: **e. Thigh.**

51 Masseter: **a. Head.**

52 Sternocleidomastoid: **d. Neck.**

53 Which of the following is *not* included in the quadriceps group? **c. Rectus abdominis.** The term *abdominis* is the giveaway here because it should make you think of the abdomen. The rest of the quadriceps group is found in the thigh, where it belongs.

54 What "two-headed" muscle is found in both the arm and the leg? **c. Biceps.** Your biggest clue: The "bi" part of biceps. Although people usually think of the biceps brachii in the arm, you can't forget about the biceps femoris at the back of the thigh.

55 This muscle divides the thoracic cavity from the abdominal cavity. **a. Diaphragm.** And without it, you couldn't breathe.

56 Together the gastrocnemius and the soleus muscles are referred to as the triceps surae. Along with a third muscle, the plantaris, they contribute to the strongest tendon in the body: **d. the Achilles tendon.** Together, the gastrocnemius (literally "stomach of the leg") and soleus ("sole fish") are commonly called the calf muscle.

57 Which of the following is *not* one of the muscles referred to as hamstrings? **b. Gracilis.** The other three answer options all are listed as hamstring muscles.

Chapter 7

It's Skin Deep: The Integumentary System

Did you know that the skin is the body's largest organ? In an average person, its 17 to 20 square feet of surface area represents 15 percent of the body's weight. Self-repairing and surprisingly durable, the skin is the first line of defense against the harmful effects of the outside world. It helps retain moisture; regulates body temperature; hosts the sense receptors for touch, pain, and heat; excretes excess salts and small amounts of waste; and even stores blood to be moved quickly to other parts of the body when needed.

Skin is jam-packed with components; it has been estimated that every square inch of skin contains 15 feet of blood vessels, 4 yards of nerves, 650 sweat glands, 100 oil glands, 1,500 sensory receptors, and more than 3 million cells with an average life span of 26 days that are constantly being replaced.

In this chapter, we peel back the surface of this most-visible organ system. We also give you plenty of opportunities to test your knowledge.

Digging Deep into Dermatology

Skin — together with hair, nails, and glands — composes the *integumentary system* (shown in Figure 7-1). The name stems from the Latin verb *integere,* which means "to cover." The relevant Greek and Latin roots include *dermato* and *cutis,* both of which mean "skin."

The skin consists of two primary parts, which we describe in the following sections: the *epidermis* and the *dermis.* (The Greek root *epi–* means "upon" or "above.") Underlying the epidermis and dermis is the *hypodermis* or *superficial fascia* (also sometimes called *subcutaneous tissue*), which acts as a foundation but is not part of the skin. Composed of areolar (porous) and adipose (fat) tissue, it anchors the skin through fibers that extend from the dermis. Underneath, the hypodermis attaches loosely to tissues and organs so that muscles can move freely. Around elbow and knee joints, the hypodermis contains fluid-filled sacs called *bursae.* The fat in the hypodermis buffers deeper tissues and acts as insulation, preventing heat loss from within the body's core. The dermis and the hypodermis are home to pressure-sensitive nerve endings called *lamellated* or *Pacinian corpuscles* that respond to a deeper poke in the skin.

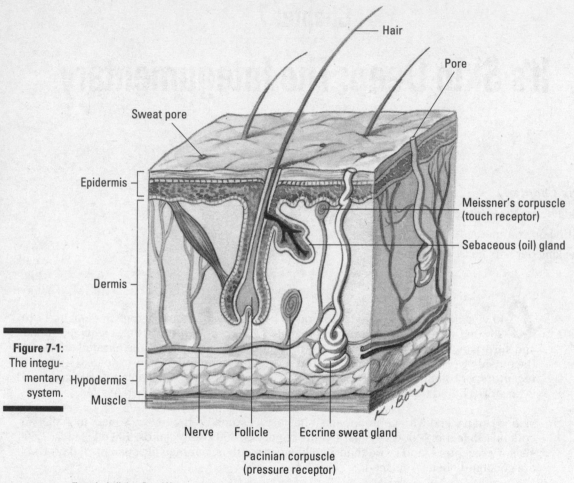

Hair

Pore

Sweat pore

Epidermis

Meissner's corpuscle
(touch receptor)

Sebaceous (oil) gland

Dermis

Figure 7-1:
The integu-
mentary
system.

Hypodermis

Muscle

Nerve   Follicle   Eccrine sweat gland

Pacinian corpuscle
(pressure receptor)

Illustration by Kathryn Born, MA

The epidermis: Don't judge this book by its cover

The epidermis, which contains no blood vessels, is made up of layers of closely packed epithelial cells. From the outside in, these layers are the following:

- **Stratum corneum** (literally the "horny layer") is about 20 layers of flat, scaly, dead cells
  containing a type of water-repellent protein called *keratin.* These cells, which represent
  about three-quarters of the thickness of the epidermis, are said to be *cornified,* which
  means that they're tough and horny like the cells that form hair or fingernails. Humans
  shed this layer of tough, durable skin at a prodigious rate; in fact, much of household
  dust consists of these flaked-off cells. Where the skin is rubbed or pressed more often,
  cell division increases, resulting in calluses and corns.

- **Stratum lucidum** (from the Latin word for "clear") is found only in the thick skin on
  the palms of the hands and the soles of the feet. This translucent layer of dead cells
  contains *eleidin,* a protein that becomes keratin as the cells migrate into the stratum
  corneum, and it consists of cells that have lost their nuclei and cytoplasm.

✔ **Stratum granulosum** is three to five layers of flattened cells containing *keratohyalin,* a substance that marks the beginning of keratin formation. No nourishment from blood vessels reaches this far into the epidermis, so cells are either dead or dying by the time they reach the stratum granulosum. The nuclei of cells found in this layer are degenerating; when the nuclei break down entirely, the cell can't metabolize nutrients and dies.

✔ **Stratum spinosum** (also sometimes called the *spinous layer*) has five to ten layers containing *prickle cells,* named for the spinelike projections that connect them with other cells in the layer. *Langerhans cells,* believed to be involved in the body's immune response, are prevalent in the upper portions of this layer and sometimes the lower part of the stratum granulosum; they migrate from the skin to the lymph nodes in response to infection. Some mitosis (cell division) takes place in the stratum spinosum, but the cells lose the ability to divide as they mature.

✔ **Stratum basale** (or *stratum germinativum*) is also referred to as the *germinal layer* because this single layer of mostly columnar stem cells generates all the cells found in the other epidermal layers. It rests on the papillary (rough or bumpy) surface of the dermis, close to the blood supply needed for nourishment and oxygen. The mitosis that constantly occurs here replenishes the skin; it takes about two weeks for the cells that originate here to migrate up to the stratum corneum, and it's another two weeks before they're shed. About a quarter of this layer's cells are *melanocytes,* cells that synthesize a pale yellow to black pigment called *melanin* that contributes to skin color and provides protection against ultraviolet radiation (the kind of radiation found in sunlight). The remaining cells in this layer become *keratinocytes,* the primary epithelial cell of the skin. Melanocytes secrete melanin directly into the keratinocytes in a process called *cytocrine secretion. Merkel cells,* a large oval cell believed to be involved in the sense of touch, occasionally appear amid the keratinocytes.

In addition to melanin, the epidermis contains a yellowish pigment called *carotene* (the same one found in carrots and sweet potatoes). Found in the stratum corneum and the fatty layers beneath the skin, it produces the yellowish hue associated with Asian ancestry or increased carrot consumption. The pink to red color of Caucasian skin is caused by *hemoglobin,* the red pigment of the blood cells. Because Caucasian skin contains relatively less melanin, hemoglobin can be seen more easily through the epidermis. Sometimes the limited melanin in Caucasian skin pools in small patches. Can you guess the name of those patches of color? Yep, they're freckles. Albinos, on the other hand, have no melanin in their skin at all, making them particularly sensitive to ultraviolet radiation.

Uplifted portions of the dermis, *dermal papillae,* cause ridges and grooves to form on the outer surface of the epidermis that increase the friction needed to grasp objects or move across slick surfaces. On hands and feet, these ridges form patterns of loops and whorls — fingerprints, palm prints, and footprints — that are unique to each person. You leave these imprints on smooth surfaces because of the oily secretions of the sweat glands on the skin's surface. In addition to these finer patterns, the areas around joints develop patterns called *flexion lines.* Deeper and more permanent lines are called *flexion creases.*

The dermis: Going more than skin deep

REMEMBER

Beneath the epidermis is a thicker, fibrous structure called the dermis, or *corium.* It consists of the following two layers, which blend together:

✔ **The outer, soft *papillary layer*** contains elastic and reticular (netlike) fibers that project into the epidermis to bring blood and nerve endings closer. *Papillae* (fingerlike projections) containing loops of capillaries increase the surface area of the dermis and anchor the epidermis. Some of these papillae contain *Meissner's corpuscles,* nerve endings that are sensitive to soft touch. It's the dermal papillae that form the epidermal ridges referred to as fingerprints.

✔ **The inner, thicker *reticular layer*** (from the Latin word *rete* for "net") is made up of dense, irregular connective tissue containing interlacing bundles of collagenous and elastic fibers that form the strong, resistant layer used to make leather and suede from animal hides. This layer is what gives skin its strength, extensibility, and elasticity. Within the reticular layer are *sebaceous glands* (oil glands), sweat glands, fat cells, hair follicles, and larger blood vessels.

Cells in the dermis include *fibroblasts* (from which connective tissue develops), *macrophages* (which engulf waste and foreign microorganisms), and adipose tissue. Thickest on the palms of the hands and soles of the feet, the dermis is thinnest over the eyelids, penis, and scrotum. It's thicker on the back (posterior) than on the stomach (anterior) and thicker on the sides (lateral) than toward the middle of the body (medial). The various skin "accessories" — blood vessels, nerves, glands, and hair follicles — are embedded here.

See if you've got the skinny on skin so far:

Q. The layer of epidermis that's composed of a horny, cornified tissue that sloughs off is called the stratum

　a. corneum.

　b. lucidum.

　c. granulosum.

　d. spinosum.

　e. basale.

A. The correct answer is corneum. Cornified, corns — think of how hard a kernel of popcorn can be.

1. Mitosis takes place in the layer of epidermis called the stratum

　a. corneum.

　b. lucidum.

　c. granulosum.

　d. papilla.

　e. basale.

2. What does the papillary layer of the dermis *not* do?

　a. Filter out microbes

　b. Extend into the epidermis

　c. Carry the blood and nerve endings close to the epidermis

　d. Aid in holding the epidermis and dermis together

　e. Contain cells sensitive to touch

3. The epidermal ridges on the fingers function to

   a. provide a means of identification.

   b. increase the friction of the epidermal surface.

   c. decrease water loss by the tissues.

   d. aid in regulating body temperature.

   e. prevent bacterial infection.

4. Caucasian skin color is caused by

   a. carotene pigment in the dermis.

   b. the high level of melanin in the epidermis.

   c. less melanin in the skin, allowing the blood pigment to be seen.

   d. the absence of all pigment

   e. melanin and carotene pigments.

5. The sequence of layers in the epidermis from the dermis outward is

   a. corneum, lucidum, granulosum, spinosum, basale.

   b. corneum, granulosum, lucidum, basale, spinosum.

   c. spinosum, basale, granulosum, corneum, lucidum.

   d. basale, spinosum, granulosum, lucidum, corneum.

   e. basale, lucidum, corneum, spinosum, granulosum.

6. What's another name for the subcutaneous layer of tissue?

   a. Epidermis

   b. Superficial fascia

   c. Papillary layer

   d. Inner reticular layer

   e. Dermis

7. Fingerprint lines are determined by

   a. the contours of the dermal papillae.

   b. the thickness of the dermis.

   c. the bundles of collagenous and elastic fibers in the epidermis.

   d. the fibroblasts, macrophages, and adipose tissue surrounding the nerves.

   e. the surface layer of cells that are constantly being shed.

8. What do carrots and sweet potatoes have in common with human epidermis?

    a. They synthesize melanocytes.

    b. They have layers of cells at varying stages of development.

    c. They contain the pigment carotene.

    d. They protect against ultraviolet radiation.

    e. They have a papillary surface.

9. What is the name of a layer of dense, irregular connective tissue containing interlacing bundles of collagenous and elastic fibers?

    a. Basale layer of the epidermis

    b. Reticular layer of the dermis

    c. Outer layer of the hypodermis

    d. Papillary layer of the dermis

    e. Inner layer of the hypodermis

10. Why is keratin important to the skin?

    a. It makes the stratum corneum thick, tough, and water-repellant.

    b. It keeps the stratum lucidum moisturized.

    c. It helps nourish the stratum granulosum.

    d. It maintains connections between the cells of the stratum spinosum.

    e. It distributes sweat evenly through the epidermal layer.

11. The stratum _____ epidermal layer contains keratohyalin.

    a. germinativum

    b. spinosum

    c. lucidum

    d. granulosum

    e. corneum

12. The layer of skin attached to the hypodermis is the

    a. papillary layer of the dermis.

    b. stratum granulosum.

    c. stratum germinativum.

    d. holorine layer of the epidermis.

    e. reticular layer of the dermis.

13. What do the prickle cells in the spinous layer do?

   a. They metabolize nutrients.

   b. They connect the cells of the layer.

   c. They transform other cells that have lost their nuclei and cytoplasm.

   d. They move nutrients from the blood into the epidermis.

   e. They secrete keratohyalin.

14. What is the name for melanin that forms into patches?

   a. Flexion creases

   b. Freckles

   c. Dermal papillae

   d. Lamellated corpuscles

   e. Matrix

Touching a Nerve in the Integumentary System

At least four kinds of receptors are involved in creating the sensation of touch:

✔ **Free nerve endings:** These are the most abundant type of sensory endings, occurring widely in the integument and within muscles, joints, viscera, and other structures. Afferent nerve endings are *dendrites* (branched extensions) of sensory neurons that act primarily as pain receptors, although some sense temperature, touch, and muscles (including the sensation of "stretch"). Found all over the body, free nerve endings are especially prevalent in epithelial and connective tissue. These small-diameter fibers have a swelling at the end that responds to touch and sometimes heat, cold, or pain. Some of these endings are specialized disc-shaped structures called *Merkel discs,* present in small numbers at the dermal-epidermal junction, that function as light-touch receptors within the deep layers of the epidermis.

✔ **Meissner's corpuscles:** These light-touch mechanoreceptors lie within the dermal papillae (you can see one in Figure 7-1). They're small, egg-shaped capsules of connective tissue surrounding a spiraled end of a dendrite. Most abundant in sensitive skin areas such as the lips and fingertips, these corpuscles and free nerve endings can sense a quick touch but not a sustained one. That's why your skin is able to ignore the touch sensation of your own clothing.

✔ **Pacinian corpuscles:** These deep-pressure mechanoreceptors are dendrites surrounded by concentric layers of connective tissue (check one out in Figure 7-1). Found deep within the dermis and hypodermis, they respond to deep or firm pressure and vibrations. Each is more than 2 millimeters long and therefore visible to the naked eye.

✔ **Hair nerve endings:** These mechanoreceptors respond to a change in position of a hair. They consist of bare dendrites. The sensory nerve enters the hair follicle, wrapping around the hair bulb. When the hair is moved, the hair stimulates the nerve endings, creating sensation.

There are two primary temperature receptors, one for heat and one for cold:

- ✔ **End-bulbs of Krause:** Also known as *Krause's corpuscles,* they contain sensory nerve endings, which may be mechanoreceptors that respond to continuous pressures. However, they are also thought to be thermoreceptors sensitive to cold and activated by temperatures below 68 degrees Fahrenheit (20 degrees Celsius). They consist of a bulbous capsule surrounding the dendrite and are commonly found throughout the body in the dermis as well as in the lips, the tongue, and the conjunctiva of the eyes.

- ✔ **Brushes of Ruffini:** Also known as *Ruffini cylinders, Ruffini's corpuscles,* or *bulbous corpuscles,* these end organs branch out parallel to the skin and detect tension deep in the skin. Thus, they are sensitive to pressure, stretching, and distortion of the skin. Found primarily in the dermis and subcutaneous tissue, they're dendrite endings enclosed in a flattened capsule.

The muscles have neuromuscular spindles (also called *proprioceptors*) that transmit information to the spinal cord and brain about the lengths and tensions of muscles. This information helps provide awareness of the body's position and the relative position of body parts. The spindles also assist with muscle coordination and muscle action efficiency.

Test whether you're staying in touch with this section:

15. Meissner's corpuscles play a role in which function?

a. The sensation of heavy pressure

b. The skin's ability to rebound into shape after pressure is applied

c. The immediate withdrawal response from intense heat

d. The sense of motion

e. The sensation of soft touch

16. In addition to pressure, what kind of stimulus can end-bulbs of Krause detect?

a. Heat above 98.6 degrees Fahrenheit

b. Cold below 68 degrees Fahrenheit

c. Stabbing sensations

d. Cold below 45 degrees Fahrenheit

e. Heat above 77 degrees Fahrenheit

Accessorizing with Hair, Nails, and Glands

Mother Nature has accessorized your fashionable over-wrap with a variety of specialized structures that grow from the epidermis: hair, fingernails, toenails, and several types of glands.

Wigging out about hair

Like most mammals, hair covers the entire human body except for the lips, eyelids, palms of the hands, soles of the feet, nipples, and portions of external reproductive organs. But human body hair generally is sparser and much lighter in color than that sported by most other mammals. Animals have hair for protection and temperature control. For humans, however, body hair is largely a secondary sex characteristic.

A thick head of hair protects the scalp from exposure to the sun's harmful rays and limits heat loss. Eyelashes block sunlight and deflect debris from the eyes. Hair in the nose and ears prevents airborne particles and insects from entering. Touch receptors connected to hair follicles respond to the lightest brush.

The average adult has about 5 million hairs, with about 100,000 of those growing from the scalp. Normal hair loss from an adult scalp is about 70 to 100 hairs each day, although baldness can result from genetic factors, hormonal imbalances, scalp injuries, disease, dietary deficiencies, radiation, or chemotherapy.

Each hair grows at an angle from a follicle embedded in the epidermis and extending into the dermis (refer to Figure 7-1); scalp hairs sometimes reach as far as the hypodermis. Nerves reach the hair at the follicle's expanded base, called the *bulb,* where a nipple-shaped papilla of connective tissue and capillaries provide nutrients to the growing hair. Epithelial cells in the bulb divide to produce the hair's *shaft* (the part that extends out of the follicle). (The part of the hair within the follicle is called the *root.*) The shape of a hair's cross section can vary from round to oval or even flat; oval hairs grow out appearing wavy or curly, flat hairs appear kinky, and round hairs grow out straight. Each scalp hair grows for two to three years at a rate of about ⅓ to ½ millimeter per day, or 10 to 18 centimeters per year. When mature, the hair rests for three or four months before slowly losing its attachment. Eventually, it falls out and is replaced by a new hair.

Hair pigment (which is melanin, just as in the skin) is produced by melanocytes in the follicle and transferred to the hair's cortex and medulla cells. Three types of melanin — black, brown, and yellow — combine in different quantities for each individual to produce different hair colors ranging from light blonde to black. Gray and white hairs grow in when melanin levels decrease and air pockets form where the pigment used to be.

Wondering why you have to shampoo so often? Hair becomes oily over time thanks to *sebum,* a mixture of cholesterol, fats, and other substances secreted from a *sebaceous* (or *holocrine*) gland found next to each follicle (see one for yourself in Figure 7-1). Sebum keeps both hair and skin soft, pliable, and waterproof. Attached to each follicle is a smooth muscle called an *arrector pili* (literally "raised hair") that both applies pressure to the sebaceous gland and straightens the hair shaft, elevating the skin in a pattern called *goose bumps* or *goose pimples.*

Each hair is made up of three concentric layers of keratinized cells:

- ✔ **The medulla:** A central core, called the *medulla,* consists of large cells containing eleidin that are separated by air spaces; in fine hair, the medulla may be small or entirely absent.

- ✔ **The cortex:** A *cortex* surrounding the medulla forms the major part of the hair shaft with several layers of flattened cells. The cortex also has elongated, pigment-bearing cells in dark hair and air pockets in white hair.

✔ **The cuticle:** The outermost *cuticle* is a single layer of overlapping cells with the free end pointing upward. The cuticle strengthens and compacts the inner layers, but abrasion tends to wear away the end of the shaft, exposing the medulla and cortex in a pattern known as *split ends*.

Nailing the fingers and toes

The hard part of the fingernails and toenails contains a tough protein called keratin. Human nails (which actually are vestigial claws) have three parts: a *root bed* at the nail base, a *body* that's attached to the fingertip, and a *free edge* that grows beyond the end of the finger or toe.

Heavily cornified tissue forms the nails from modified strata corneum and lucidum (we describe these and other layers in the earlier section "The epidermis: Don't judge this book by its cover"). A narrow fold of the stratum corneum turns back to form the *eponychium,* or *cuticle.* Under the nail, the nail bed is formed by the strata basale and spinosum. At the base of the nail, partially tucked under the cuticle, the strata thicken to form a whitish area called the *lunula* (literally "little moon") that can be seen through the nail. Beneath the lunula is the *nail matrix,* a region of thickened strata where mitosis pushes previously formed cornified cells forward, making the nail grow. Under the free edge of the nail, the stratum corneum thickens to form the *hyponychium.* Nails are pinkish in color because of hemoglobin in the underlying capillaries, which are visible through the translucent cells of the nail.

On average, fingernails grow about 1 millimeter each week. Toenails tend to grow even more slowly. Nails function as an aid to grasping, as a tool for manipulating small objects, and as protection against trauma to the ends of fingers and toes.

Sweating the details

Humans perspire over nearly every inch of skin, but anyone with sweaty palms or smelly feet can attest to the fact that sweat glands are most numerous in the palms and soles, with the forehead running a close third. There are two types of sweat, or *sudoriferous,* glands: *eccrine* and *apocrine.* Both are coiled tubules embedded in the dermis or subcutaneous layer composed of simple columnar cells (refer to Figure 7-1).

✔ **Eccrine sweat glands** are distributed widely over the body — an average adult has roughly 3 million of them — and produce the watery, salty secretion you know as sweat. Each gland's duct passes through the epidermis to the skin's surface, where it opens as a *sweat pore.* The sympathetic division of the autonomic nervous system (see Chapter 15) controls when and how much perspiration is secreted depending on how hot the body becomes. Sweat helps cool the skin's surface by evaporating as fast as it forms. About 99 percent of eccrine-type sweat is water, but the remaining 1 percent is a mixture of sodium chloride and other salts, uric acid, urea, amino acids, ammonia, sugar, lactic acid, and ascorbic acid.

✔ **Apocrine sweat glands** are located primarily in armpits (known as the *axillary region*) and the external genital area. Usually associated with hair follicles, they produce a white, cloudy secretion that contains organic matter. Although apocrine-type sweat contains the same basic components as eccrine sweat and also is odorless when first secreted, bacteria quickly begin to break down its additional fatty acids and proteins — explaining the post-exercise underarm stench. In addition to exercise, sexual and other emotional stimuli can cause contraction of cells around these glands, releasing sweat.

Getting an earful

The occasionally troublesome yellowish substance known as earwax is secreted in the outer part of the ear canal from modified sudoriferous glands called *ceruminous glands* (the Latin word *cera* means "wax"). Lying within the subcutaneous layer of the ear canal, these glands have ducts that either open directly into the ear canal or empty into the ducts of nearby sebaceous glands, mixing their secretion with sebum and dead epithelial cells. Technically called *cerumen,* earwax is the combined secretion of these two glands. Working with ear hairs, cerumen traps any foreign particles before they reach the eardrum. As the cerumen dries, it flakes and falls from the ear, carrying particles out of the ear canal.

Think you've got a grip on everything to do with hair, nails, and glands? Find out by answering the following practice questions:

17. What is another name for the cuticle?

   a. Lunula

   b. Hyponychium

   c. Eponychium

   d. Nail matrix

   e. Perinychium

18. The _____ glands form perspiration.

   a. sebaceous

   b. ceruminous

   c. endocrine

   d. Merkel

   e. sudoriferous

19. The cause of graying hair is

   a. production of melanin in the shaft of the hair.

   b. production of carotene in the shaft of the hair.

   c. decrease in blood supply to the hair.

   d. lack of melanin in the shaft of the hair.

   e. parenthood.

20. From where does the hair develop?

   a. Arrector pili

   b. Shaft

   c. Follicle

   d. Sebaceous gland

   e. Lanugo

21. The nails are modifications of the epidermal layers

 a. corneum and lucidum.

 b. lucidum and granulosum.

 c. granulosum and spinosum.

 d. spinosum and basale.

 e. lucidum and spinosum.

22. What is the name of the muscle that straightens a hair and puts pressure on a gland causing it to secrete sebum?

 a. Terminalis muscle

 b. Arrector pili muscle

 c. Internal oblique muscle

 d. External transversus muscle

 e. Internal rectus muscle

23. Which factor is not associated with baldness?

 a. Genetics

 b. Hormonal imbalances

 c. Scalp injuries

 d. Lack of carotene

 e. Disease

24. Sebaceous glands

 a. produce a watery solution called sweat.

 b. produce an oily mixture of cholesterol, fats, and other substances.

 c. produce a waxy secretion called cerumen.

 d. accelerate aging.

 e. are associated with endocrine glands.

25. What function does the bulb at the base of a hair follicle serve?

 a. To prevent dirt and debris from becoming embedded in the skin

 b. To establish additional thermal protection

 c. To provide nutrients to the growing hair

 d. To regulate sweat production

 e. To inject melanin into the hair

26. These glands aid in cooling of the skin's surface.

   a. Endocrine glands

   b. Sebaceous glands

   c. Ceruminous glands

   d. Prickle glands

   e. Eccrine sweat glands

27. This gland contains true sweat, fatty acids, and proteins and acquires an unpleasant odor when bacteria breaks down the organic molecules it secretes.

   a. Apocrine sweat gland

   b. Sebaceous gland

   c. Ceruminous gland

   d. Eccrine sweat gland

   e. Mammary gland

28. Which of the following is true about fingernails?

   a. They're derived from the hypodermis.

   b. They contain carotene.

   c. They grow more slowly than toenails.

   d. They grow about 1 millimeter per week.

   e. They're not a skin accessory.

Answers to Questions on the Skin

The following are answers to the practice questions presented in this chapter.

1 Mitosis takes place in the layer of epidermis called the stratum **e. basale.**

This layer also is called the stratum germinativum, but a simpler memory tool is to associate it with the "base" of the epidermis.

2 What does the papillary layer of the dermis *not* do? **a. Filter out microbes.** Busy little fingerlike projections, those papillae, but they're not filters.

3 The epidermal ridges on the fingers function to **b. increase the friction of the epidermal surface.** They're Mother Nature's way of helping you cling to tree branches or grab food.

4 Caucasian skin color is caused by **c. less melanin in the skin, allowing the blood pigment to be seen.** Here's a fun experiment: Turn off the lights, press your fingers together, and hold a flashlight under them. See the red glow? That's hemoglobin, too.

5 The sequence of layers in the epidermis from the dermis outward is **d. basale, spinosum, granulosum, lucidum, corneum.**

Memory tool time: Base, spine, grain, Lucy, corny. Or try the first letters of Be Super Greedy, Less Caring. Insensitive, yes, but effective.

6 What's another name for the subcutaneous layer of tissue? **b. Superficial fascia.** Subcutaneous is the same as hypodermis (from the Greek *hypo–* for "beneath").

7 Fingerprint lines are determined by **a. the contours of the dermal papillae.** Those papillae also serve to increase the surface area of the dermis and anchor the epidermis.

8 What do carrots and sweet potatoes have in common with human epidermis? **c. They contain the pigment carotene.** Most of the other answers wouldn't apply to both mammalian and vegetable cells.

9 What is the name of a layer of dense, irregular connective tissue containing interlacing bundles of collagenous and elastic fibers? **b. Reticular layer of the dermis.** The description in this question sounds like a tough structure, so it may help you to remember that the reticular layer is what's used to make leather from animal hides.

10 Why is keratin important to the skin? **a. It makes the stratum corneum thick, tough, and water-repellant.** Associate the words "corneum" and "keratin," and you're in great shape.

11 The stratum **d. granulosum** epidermal layer contains keratohyalin. Keratohyalin eventually becomes keratin, so think of the layer where the cells are starting to die off.

12 The layer of skin attached to the hypodermis is the **e. reticular layer of the dermis.** Reticular means netlike; it makes sense that this netting lies between the dermis and the hypodermis.

13 What do the prickle cells in the spinous layer do? **b. They connect the cells of the layer.** The spine-like projections that make those connections also make the cells look prickly, hence the name.

14 What is the name for melanin that forms into patches? **b. Freckles.** Ever noticed how kids have more freckles at the end of a long summer spent outdoors? That's ultraviolet radiation working on those melanin patches.

15 Meissner's corpuscles play a role in which function? **e. The sensation of soft touch.** Although it's true that several different nerves are involved in the overall sense of touch, Meissner's corpuscles are the most responsive to touch.

16 In addition to pressure, what kind of stimulus can end-bulbs of Krause detect? **b. Cold below 68 degrees Fahrenheit.**

 Specific temperatures may seem tough to remember, but look at it this way: When it's 45 degrees Fahrenheit, you definitely need a jacket. When it's 77 degrees Fahrenheit, you don't. But when it's 68 degrees Fahrenheit, you'll want to carry a light jacket in case it gets colder. Voilà! A cold receptor activation temperature!

17 What is another name for the cuticle? **c. Eponychium.** Recall that the prefix *ep–* refers to "upon" or "around," whereas the prefix *hypo–* refers to "below" or "under." The cuticle is around the base of the nail, so it's the eponychium, not the hyponychium.

18 The **e. sudoriferous** glands form perspiration. The Latin word *sudor* means "sweat."

19 The cause of graying hair is **d. lack of melanin in the shaft of the hair.** Despite the medical cause, people often suspect that answer "e. parenthood" has a lot to do with graying hair.

20 From where does the hair develop? **c. Follicle.** The Latin translation of this word is "small cavity" or "sac," so it makes sense that this would be an origination place.

21 The nails are modifications of the epidermal layers **a. corneum and lucidum.** These are the two upper layers.

22 What is the name of the muscle that straightens a hair and puts pressure on a gland causing it to secrete sebum? **b. Arrector pili muscle.** *Arrector* is similar to "erector" (in fact, the muscle is sometimes called that), which implies straightening, and the Latin word for "hair" is *pili.*

23 Which factor is not associated with baldness? **d. Lack of carotene.** This answer just means that your hair won't turn orange, not necessarily that it will fall out of your scalp.

24 Sebaceous glands **b. produce an oily mixture of cholesterol, fats, and other substances.** That secretion's called *sebum* — hence "sebaceous glands."

25 What function does the bulb at the base of a hair follicle serve? **c. To provide nutrients to the growing hair.** This is where connective tissue and capillaries come together to provide those nutrients.

26 These glands aid in cooling of the skin's surface. **e. Eccrine sweat glands.** The evaporation of perspiration cools the skin naturally.

27 This gland contains true sweat, fatty acids, and proteins and acquires an unpleasant odor when bacteria breaks down the organic molecules it secretes. **a. Apocrine sweat gland.** These are the truly stinky sweat glands.

 Here's a memory tool for the difference between apocrine and eccrine sweat glands: You may have to APOlogize for your APOcrine glands but not your eccrine glands.

28 Which of the following is true about fingernails? **d. They grow about 1 millimeter per week.** The other answer options are either just plain wrong or nonsensical.

Part III
Feed and Fuel: Supply and Transport

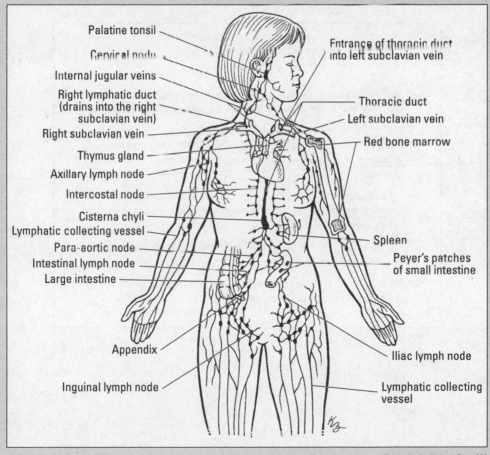

Palatine tonsil

Cervical node

Internal jugular veins

Right lymphatic duct
(drains into the right
subclavian vein)

Right subclavian vein

Thymus gland

Axillary lymph node

Intercostal node

Cisterna chyli

Lymphatic collecting vessel

Para-aortic node

Intestinal lymph node

Large intestine

Appendix

Inguinal lymph node

Entrance of thoracic duct
into left subclavian vein

Thoracic duct

Left subclavian vein

Red bone marrow

Spleen

Peyer's patches
of small intestine

Iliac lymph node

Lymphatic collecting
vessel

Illustration by Kathryn Born, MA

Research in what makes the human body tick is going in entirely new directions with study of the human microbiome. Catch the latest at www.dummies.com/extras/anatomyphysiologywb.

In this part . . .

✔ Breathe in the details of the respiratory system, including the parts of the respiratory tract (the nose, the sinuses, the throat, and the lungs).

✔ Fuel up with food and then study what happens to it in the digestive system after every possible nutrient has been wrung from it.

✔ Take the pulse of the circulatory system and discover how its internal transit routes carry nutrients and oxygen to every nook and cranny of the body.

✔ Delve into the lymphatic ducts and find out about the body's internal defense forces.

✔ Bat clean-up with the kidneys and track how they move waste from the body as part of the urinary system.

Chapter 8

Oxygenating the Machine: The Respiratory System

In This Chapter

▶ Tracking respiration: In with oxygen, out with carbon dioxide

▶ Identifying the organs and muscles of the respiratory tract

▶ Taking note of common pulmonary diseases

People need lots of things to survive, but the most urgent need from moment to moment is oxygen. Without a continual supply of this vital element, we don't last long. But if we have reserves of the other things we need — carbohydrates, fats, and proteins — why don't we have some kind of storehouse of oxygen, too? Simple. It's readily available in the air around us, so we've never needed to evolve a means for storing it. Nonetheless, our stored food supplies would be useless without oxygen; our bodies can't metabolize the energy they need from these substances without a constant stream of oxygen to keep things percolating along.

All that metabolizing creates another equally important need, however. We must have a means for getting rid of our bodies' key gaseous waste — carbon dioxide, or CO_2. If it builds up in our systems, we die. It must be removed from our bodies almost as fast as it's formed. Conveniently, breathing in fulfills our need for oxygen and breathing out fulfills our need to expel carbon dioxide.

In this chapter, you get a quick review of Mother Nature's dual-purpose system and plenty of opportunities to test your knowledge about the lungs and other parts of the respiratory system.

Breathing In Oxygen, Breathing Out CO₂

Respiration, or the exchange of gases between an organism and its environment, occurs in three distinct processes:

> ✔ **Breathing:** The technical term is *pulmonary ventilation,* or the movement of air into and out of the lungs. (Breathing is also called *inspiration* and *expiration.*)
>
> ✔ **Exchanging gases:** This takes place between the alveolar cells in the lungs, the blood, and the body's cells in two ways:
>
> > • **Pulmonary, or external, respiration:** The exchange in the lungs when blood gains oxygen and loses carbon dioxide, transforming it from venous blood (deoxygenated) into arterial blood (oxygenated)

- **Systemic, or internal, respiration:** The exchange that takes place in and out of capillaries when the blood releases some of its oxygen and collects carbon dioxide from the tissues

✔ **Cellular respiration:** Oxygen is used in the catabolism of substances like glucose for the production of energy (see Chapter 1).

Here are some key respiratory terms to keep in mind:

✔ **Adult breathing rate:** About 12 to 20 times per minute.

✔ **Anoxia:** Oxygen deficiency in which the cells either don't have or can't utilize sufficient oxygen to perform normal functions.

✔ **Asphyxia:** Lack of oxygen with an increase in carbon dioxide in the blood and tissues; accompanied by a feeling of suffocation leading to coma.

✔ **Expiration or exhalation:** The diaphragm returns to its domed shape as the muscle fibers relax, via elastic recoil of the lungs and tissues lining the thoracic cavity; the external intercostal muscles relax; and the internal intercostal muscles contract. This movement pulls the ribs back into place, decreasing the volume of the thoracic cavity and increasing pressure, forcing air out of the lungs.

✔ **Hypoxia:** Low oxygen content in the inspired air. This occurs naturally at high altitudes and may result in altitude sickness among hikers and pilots.

✔ **Inspiration or inhalation:** When the muscles of the diaphragm contract, its dome shape flattens; simultaneously, the contraction of the external intercostal muscles pulls the ribs upward and increases the volume of the thoracic cavity, decreasing the intra-alveolar pressure. The pressure difference between the atmosphere and the lungs diffuses air into the respiratory tract.

✔ **Lung capacity:** The vital capacity plus the residual air volume.

✔ **Mediastinum:** The region between the lungs extending from the sternum *ventrally* (at the front) to the thoracic vertebrae *dorsally* (at the back), and *superiorly* (top) from the entrance of the thoracic cavity to the diaphragm *inferiorly* (at the bottom).

✔ **Minimal air:** The volume of air in the lungs when they're completely collapsed (150 cubic centimeters in an adult).

✔ **Phrenic nerve:** The nerve that *innervates* (stimulates) the diaphragm.

✔ **Residual air volume:** The volume of air remaining in the lungs after the most forceful expiration (1,200 cubic centimeters in an adult).

✔ **Respiratory centers:** Nerve centers for regulating breathing located in the *medulla oblongata and pons* of the brain stem. The centers are influenced by the amount of carbon dioxide in the blood.

✔ **Tidal air:** The volume of air inspired and expired in the resting state (500 cubic centimeters in an adult).

✔ **Vital capacity:** The volume of air moved by the most forceful expiration after a maximum inspiration. It represents the total moveable air in the lungs (4,600 cubic centimeters in an adult).

Here's what happens as you breathe in and out (see Figure 8-1): Red blood cells use a pigment called *hemoglobin* to carry oxygen and carbon dioxide simultaneously throughout the body through the circulatory system (for more on that system, turn to Chapter 10). Hemoglobin bonds loosely with oxygen, or O_2, to carry it; the bonded hemoglobin is called *oxyhemoglobin*.

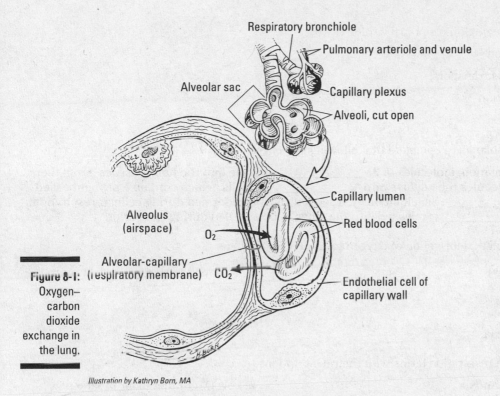

Respiratory bronchiole

Pulmonary arteriole and venule

Alveolar sac

Capillary plexus

Alveoli, cut open

Capillary lumen

Alveolus (airspace)

O_2

Red blood cells

Alveolar-capillary (respiratory membrane)

CO_2

Endothelial cell of capillary wall

Figure 8-1: Oxygen–carbon dioxide exchange in the lung.

Illustration by Kathryn Born, MA

After hemoglobin releases its oxygen molecules, it picks up carbon dioxide, or CO_2, to deliver to the lungs for exhalation. The freshly bonded hemoglobin becomes *carbohemoglobin (carbhemoglobin* or *carbaminohemoglobin).*

See whether you're carrying away enough information about respiration by tackling the following practice questions:

EXAMPLE

Q. The air that moves in and out of the lungs during normal, quiet breathing is called

a. tidal volume, or tidal air.

b. inspiratory reserve.

c. vital capacity.

d. lung capacity.

A. The correct answer is tidal volume, or tidal air. The question asks only about air moved during normal, quiet breathing, not the kind of forceful air movement involved in measuring lung capacity. Think of the normal ebb and flow of the ocean's tide as opposed to the waves of a raging storm.

1. What does the mediastinum have to do with the lungs?

a. It's the bone that protects them from collapsing during blunt trauma.

b. It's the region between them.

c. It's the control mechanism that reduces minimal air when the body is stressed.

d. It's the muscle structure that moves the diaphragm to inflate and deflate them.

2. What is hemoglobin's primary role?

   a. Filtration

   b. Gaseous transport

   c. Ventilation

   d. Asphyxia

3.–6. Fill in the blanks to complete the following sentences:

Upon inhalation, molecules of **3.** _____ diffuse into the lung's tissues. From there, these molecules then diffuse into **4.** _____ cells, which contain a pigment called **5.** _____. Simultaneously, a second substance formed during cellular respiration, **6.** _____, is released into the lungs to be expelled during exhalation.

7. The complete deprivation of oxygen to tissues is known as

   a. asphyxia.

   b. anoxia.

   c. hypoxia.

   d. catapoxia.

8.–12. Match the respiration terms with their descriptions.

8. _____ Anoxia	**a.** Low O_2 content in air breathed in
9. _____ Systemic or internal respiration	**b.** O_2 lacking and excessive CO_2
10. _____ Asphyxia	**c.** Gaseous exchange between capillaries and cells
11. _____ Hypoxia	**d.** Without O_2 at the cell level
12. _____ Pulmonary, or external, respiration	**e.** Gaseous exchange in lungs

Inhaling the Basics about the Respiratory Tract

We fill and empty our lungs by contracting and relaxing the respiratory muscles, which include the dome-shaped diaphragm and the intercostal muscles that surround the rib cage. As these muscles contract, air moves through a series of interconnected chambers in the following order: nose → pharynx → larynx → trachea → bronchi → bronchioles → alveolar ducts → alveoli. We look at the details of each chamber in that order.

Knowing about the nose (and sinuses)

You may care a great deal about how your nose is shaped, but the shape actually makes little difference to your body. The nose is simply the most visible part of your respiratory tract. Beyond those oh-so-familiar nostrils — which are formally called *external nares* — the *septum* divides the nasal cavity into two chambers called the *nasal fossae*. The internal openings at the posterior of the fossae are called the *choanae* or *internal nares*. Inside the nostril is a slight dilation extending to the apex of the nose called the *vestibule;* it's lined by skin covered with hairs, plus mucous glands and sebaceous glands that help trap dust and particles before they can enter the lungs.

Each nasal cavity is divided into an *olfactory region* and a *respiratory region*.

 ✔ **Olfactory region:** The olfactory region lies in the upper part of the nasal cavity. Fine filaments distributed over its mucous membrane are actually special nerves devoted to the sense of smell. The bipolar olfactory cells' axons thread through openings in the *cribriform plate* (from the Latin *cribrum* for "sieve"; see Chapter 5) and then come together in the *olfactory bulb,* which acts as an odorant classification filter (for perception of smells) whose output signals are sent to the higher olfactory centers of the brain's cerebral cortex for further discrimination and enhancement of the olfactory process. Impulses flow from the olfactory bulb through the olfactory tract to two main destinations:

 • Via the *thalamus* to the *olfactory cortex,* where smells are consciously detected

 • Via the sub-cortical route to the *hypothalamus* and to other regions of the limbic systems where emotional aspects of the smells are analyzed

 ✔ **Respiratory region:** The nasal cavity's respiratory region is covered by a mucous membrane made up of pseudostratified, ciliated columnar epithelium (check out the ten types of epithelium in Chapter 4) containing mucous and sebaceous glands. The secretions from these glands form a protective layer that warms, moistens, and helps to filter air as it's inhaled. Beneath the protective layer, areolar connective tissue (discussed in Chapter 4) contains *lymphocytes* (forming a thin *lymphoid tissue*) that remove foreign materials. A layer of blood vessels next to the *periosteum* (the membrane covering the surface of bones) forms a rich *plexus* (network) that tends to swell when irritated or inflamed, closing the *ostia* (openings) of the nasal sinuses. Don't you just hate it when that happens?

Ah, the sinuses. They can be such headaches. Lined with a pseudostratified, ciliated columnar epithelium, *paranasal sinuses* are air-filled cavities in the bone that reduce the skull's weight and act as resonators for the voice. Each of the four groups of sinuses is named for the bone containing it (see Chapter 5), as follows:

 ✔ **Frontal sinuses** are located in the front bone behind the eyebrows. (If you've ever flown with a sinus infection in an airplane, these are the suckers that hit you right behind the eyes.)

 ✔ **Maxillary sinuses** are located in the *maxillae,* or upper jawbone.

 ✔ **Ethmoid** and **sphenoid sinuses** are located in the ethmoid and sphenoid bones in the cranial cavity's floor.

Beyond the sinuses and connected to them are *nasal ducts* that extend from the medial angle of the eyes to the *nasal cavity.* These tear ducts let serous fluid from the eyes' *lacrimal glands* flow into the nasal cavity.

The nasal cavity performs several important functions:

 ✔ It drains mucous secretions from the sinuses.

 ✔ It drains lacrimal secretions from the eyes.

 ✔ It prepares inhaled air for the lungs by warming, moistening, and filtering it. Dust and bacteria are caught in the mucous and passed outward from the nasal cavity by the motion of the cilia. Some of that gunk is taken up by *lymphatic tissue* in the nasal cavity and respiratory tubes for delivery to the lymph nodes, which destroy invading germs.

Beyond the nasal cavity is the *nasopharynx,* which connects — you guessed it — the nasal cavity to the *pharynx.*

With a bit of a refresher on the nasal and sinus passages, do you think you can hit the following practice questions on the nose?

13. Which of the following statements about the mucous membranes of the nasal cavity is not true?

   a. They contain an abundant blood supply.

   b. They moisten the air that flows over them.

   c. They're composed of stratified squamous epithelium.

   d. They become inflamed, causing the membrane to swell and close the nasal sinuses.

14. Why does the nose have sebaceous glands?

   a. Their secretions smooth air flow into the nostrils.

   b. They enhance the sense of smell.

   c. Their secretions form a protective layer to warm, moisten, and help filter air.

   d. They improve resonation during speech.

15.–30. Use the terms that follow to identify the structures of the respiratory tract shown in Figure 8-2.

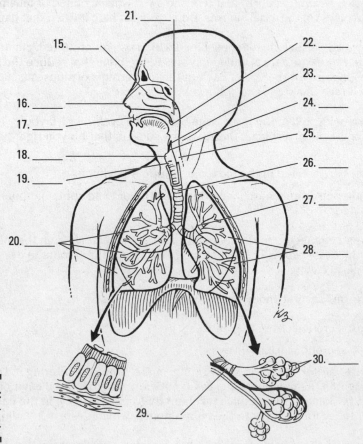

Figure 8-2:
The respiratory tract.

Illustration by Kathryn Born, MA

a. Esophagus

b. Larynx

c. Nasal cavity

d. Oropharynx

e. Nose

f. Right lung

g. Epiglottis

h. Mouth

i. Alveoli

j. Nasopharynx

k. Thyroid cartilage

l. Left lung

m. Trachea

n. Bronchioles

o. Laryngopharynx

p. Left primary bronchus

Dealing with throaty matters

In laymen's terms, it's the throat. But you know better, right? The top of the throat consists of these key parts:

- **Pharynx:** The pharynx is an oval, fibromuscular tube about 5 inches long and tapering to ½ inch in diameter at its *anteroposterior end,* which is a fancy biology term meaning "front to back." In fact, the point where the pharynx connects to the esophagus is the narrowest part of the entire digestive tract. The pharynx consists of three main divisions:

  - The *nasopharynx,* where the nasal cavity opens into the pharynx

  - The *oropharynx,* where the oral cavity opens into the pharynx

  - The *laryngopharynx,* which begins at the epiglottis and leads down to the esophagus

  Eustachian tubes connected to the middle ears enter the nasopharynx on each side. (And when the tubes become blocked by an infection, sudden pressure changes can be extremely painful). On the back wall of the nasopharynx is a mass of lymphoid tissue called the *pharyngeal tonsil,* or *adenoids.*

- **Larynx:** Connecting the pharynx with the trachea, this collection of nine cartilages is what makes a man's prominent Adam's apple. Also called the voice box, the larynx looks like a triangular box flattened dorsally and at the sides that becomes narrow and cylindrical toward the base. Ligaments connect the cartilages controlled by several muscles; the inside of the larynx is lined with a mucous membrane that continues into the trachea.

Three of the larynx's nine cartilages go solo — the *thyroid,* the *cricoid,* and the *epiglottis* — while the others come in pairs — the *arytenoids,* the *corniculates,* and the *cuneiforms.*

✔ **Thyroid cartilage:** The thyroid cartilage (*thyroid* in Greek means "shield-shaped") is largest and consists of two plates called *laminae* that are fused just beneath the skin to form a shield-shaped process, the Adam's apple. Immediately above the Adam's apple, the laminae are separated by a V-shaped notch called the superior thyroid notch.

✔ **Cricoid cartilage:** The ring-shaped cricoid cartilage is smaller but thicker and stronger, with shallow notches at the top of its broad back that connect, or articulate, with the base of the arytenoid cartilages.

✔ **Arytenoid cartilages:** The arytenoid cartilages both are shaped like pyramids, with the vocal folds attached at the back and the controlling arytenoid muscles that move the arytenoids attached at the sides, moving the vocal cords.

✔ **Corniculate cartilages:** On top of the arytenoids are the corniculate cartilages, small conical structures for attachment of muscles regulating tension on the vocal cords.

✔ **Cuneiform cartilages:** Nestled in front of these and inside the *aryepiglottic* fold, the cuneiform cartilages stiffen the soft tissues in the vicinity.

✔ **Epiglottis cartilage:** The epiglottis, sometimes called the lid on the voice box, is a leaf-shaped cartilage, attached at its stem end, that projects upward behind the root of the tongue. It opens during respiration and reflexively closes during swallowing to keep food and liquids from getting into the respiratory tract. It's composed of elastic cartilage having a flexible attachment to the anterior and superior borders of the *hyoid bone* and larynx.

Two types of folds play different roles inside the larynx.

✔ The true vocal folds, or cords, are V-shaped when relaxed. When talking, the folds stretch for high sounds or slacken for low sounds, causing the glottis — the opening in the larynx — to form an oval. Sounds are produced when air is forced over the folds, causing them to vibrate.

✔ Just above these folds are the ventricular vocal folds, also known as vestibular or false folds, that don't produce sounds. Muscle fibers within these folds help close the glottis during swallowing.

Following are some practice questions dealing with the throat:

31. What happens when the arytenoids move?

a. The Adam's apple moves.

b. The cricoid cartilage flexes.

c. The vocal cords move.

d. The cuneiform cartilages stiffen.

32.–36. Match the anatomical structure with its function.

32. _____ Epiglottis	**a.** Voice box lid
33. _____ Nasal fossa	**b.** Respiratory center
34. _____ Medulla oblongata	**c.** Prevent collapse of trachea
35. _____ C-shaped cartilaginous rings	**d.** Nasal cavity
36. _____ Thyroid cartilage	**e.** Adam's apple

37. The _____ is the opening between the two vocal folds, and the _____ covers that opening as needed.

    a. epiglottis, glottis

    b. bronchi, epiglottis

    c. alveoli, pharynx

    d. glottis, epiglottis

38. Why can it be painful to fly with a sinus infection?

    a. The infection can inflame the eardrum and cause it to bulge.

    b. The infection can block the Eustachian tube, causing unequal air pressure on each side of the eardrum.

    c. A build-up of mucous inside a sinus cavity presses against facial nerves.

    d. The infection can inflame the optic nerves.

39.–52. Use the terms that follow to identify the structures of the larynx shown in Figure 8-3. Some terms may be used more than once.

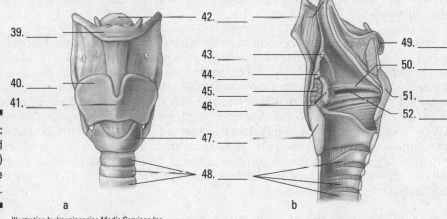

Figure 8-3: Front (a) and lateral (b) views of the larynx.

a   b

Illustration by Imagineering Media Services Inc.

    a. Thyroid cartilage

    b. Cricoid cartilage

    c. Hyoid bone

    d. Epiglottis

    e. Arytenoid cartilage

    f. Ventricular fold (false vocal cord)

    g. Laryngeal prominence (Adam's apple)

h. Cuneiform cartilage

i. Arytenoid muscle

j. True vocal cord

k. C-shaped tracheal cartilages

l. Corniculate cartilage

Going deep inside the lungs

After the pharynx and larynx comes the *trachea,* more popularly known as the windpipe. Roughly 6 inches long in adults, it's a tube connected to the larynx in front of the esophagus that's made up of C-shaped rings of hyaline cartilage and fibrous connective tissue that strengthen it and keep it open. Like the larynx, the trachea is lined with mucous membrane covered in cilia. Just above the heart, the trachea splits into two *bronchi* divided by a sharp ridge called the *carina,* with each leading to a lung. But they're not identical: The right *primary bronchus* is shorter and wider than the left primary bronchus. Each primary bronchus divides into secondary bronchi with a branch going to each lobe of the lung; the right side gets three secondary bronchi while the left gets only two. Once inside a designated lobe, the bronchus divides again into tertiary bronchi. The right lung has ten such branches: three in the superior (or upper) lobe, two in the middle lobe, and five in the inferior (or lower) lobe. The left lung has only four tertiary bronchi: two in the upper lobe and two in the lower lobe.

Each tertiary bronchus subdivides one more time into smaller tubes called *bronchioles,* which lack the supporting cartilage of the larger structures. The lining loses its cilia and changes to cuboidal and then to squamous epithelium as it extends away from its point of origin. Each bronchiole divides into alveolar ducts that end in an elongated sac called the *atrium,* which in turn ends in a structure known as an *alveolar sac.* Surrounding the atria are *alveoli* (or air sacs) and small *capillaries* from which excess carbon dioxide diffuses out and into which oxygen diffuses. Oxygen is then transported to all parts of the body. Each lung contains approximately 150 million alveoli. Overall, there are 23 generations of branching in the respiratory system forming a *fractal structure* (a geometric pattern that is repeated at ever smaller scales) with a combined surface area (counting the alveoli) the size of one side of a tennis court!

Knowing that the bronchi aren't evenly distributed, you may have guessed that the lungs aren't identical either. You're right. They both have a spongy and porous texture, but the right lung is larger, wider, and shorter than the left lung and has three *lobes.* The left lung divides into only two lobes and is both narrower and longer to make room for the heart because two-thirds of that organ lies to the left of the body's midline. Each lobe is made up of many *lobules,* each with a bronchiole ending in an atrium inside.

Covering each lung is a thin serous membrane called the *visceral pleura* that folds back on itself to form a second outer layer, the *parietal pleura,* with a *pleural cavity* between the two layers. These two layers secrete a watery fluid into the cavity to lubricate the surfaces that rub against each other as you breathe. When the pleural membrane becomes inflamed in a condition called *pleurisy,* a sticky discharge roughens the pleura, causing painful irritation. An accompanying bacterial infection means that pus accumulates in the pleural cavity in a condition known as *empyema.*

REMEMBER

Blood comes to the lungs through two sources: the *pulmonary arteries* and the *bronchial arteries.* The pulmonary trunk comes from the right ventricle of the heart and then branches into the two pulmonary arteries carrying *venous blood* (the only arteries that contain blood loaded with carbon dioxide) from various parts of the body to the lungs (see Chapter 10 for an introduction to the circulatory system). That blood goes through pulmonary arterioles and then through pulmonary capillaries in the lungs where the carbon dioxide leaves the blood and enters the alveoli to be expelled during exhalation; oxygen leaves the alveoli through the capillaries to enter the bloodstream. After that, oxygenated arterial blood leaves the lung through pulmonary venules returning to the left atrium through the *pulmonary veins* (the only veins that contain oxygenated blood), completing the cycle. Bronchial arteries branch off the thoracic aorta of the heart, supplying the lung tissue with nutrients and oxygen.

Following are some practice questions dealing with the lungs:

53. In which part of the respiratory system is cartilage never found?

   a. Primary bronchi

   b. Secondary bronchi

   c. Bronchioles

   d. Trachea

54. The capillaries and the _____ exchange gases.

   a. trachea

   b. alveolar sacs

   c. primary bronchi

   d. terminal bronchioles

55.–59. Fill in the blanks to complete the following sentences:

The trachea divides into two **55.** _____, which then divide into **56.** _____ with a branch going to each lobe of the lung. Upon entering the lobe, each divides into **57.** _____, subdividing into smaller tubes called **58.** _____. They terminate in an elongated sac called the atrium surrounded by **59.** _____ or air sacs.

60. If a pin were to pierce the body from the outside in the thoracic region, which three structures — in order — would it pass through?

   a. Lung, pleural cavity, terminal bronchioles

   b. Visceral pleura, lung, pleural cavity

   c. Parietal pleura, pleural cavity, visceral pleura

   d. Primary bronchi, secondary bronchi, terminal bronchioles

61.–67. Use the terms that follow to identify the structures of the bronchiole shown in Figure 8-4.

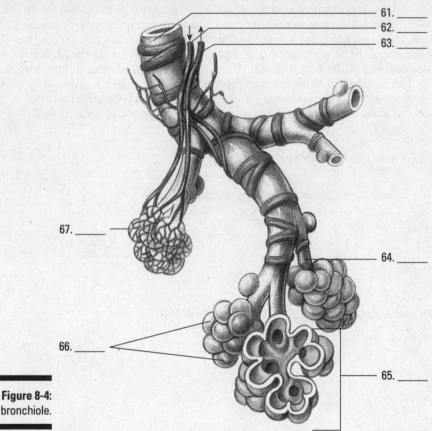

61. _____
62. _____
63. _____
67. _____
64. _____
66. _____
65. _____

Figure 8-4:
A bronchiole.

Illustration by Kathryn Born, MA

a. Pulmonary venule

b. Alveolar sac

c. Tertiary bronchiole

d. Pulmonary capillary

e. Pulmonary arteriole

f. Alveolar duct

g. Alveoli

Damaging Air

A number of pulmonary diseases can plague human lungs. Inhaling metal and mineral dust can be particularly harmful because the particles cut into and embed themselves in delicate lung tissue, leaving nonfunctional and less pliable scar tissue. Specific lung conditions include

✔ **Silicosis:** Commonly found among construction workers, silicosis is caused by deposits of sand particles in the lungs.

✔ **Anthracosis:** Also known as black lung, anthracosis occurs among coal miners because of coal dust accumulating in the lungs.

✔ **Rhinitis:** The common cold, rhinitis, can be caused by several different kinds of viral infections. Undue exposure may activate the virus or cause the body to become more susceptible to the virus.

68. Why do silicosis and anthracosis cause permanent damage to the lungs?

a. Inflammation blocks the terminal bronchioles.

b. Sand and coal dust are biodegradable.

c. Small particles cause scars inside the lungs, and scar tissue isn't as elastic as normal tissue.

d. They lead to increasing frequency of rhinitis later in life.

Answers to Questions on the Respiratory System

The following are answers to the practice questions presented in this chapter.

1 What does the mediastinum have to do with the lungs? **b. It's the region between them.** Why go with a more complex answer when the least complex is the right one?

2 What is hemoglobin's primary role? **b. Gaseous transport.** Hemoglobin is the protein that bonds with and releases gasses within the bloodstream. That's how oxygen gets where it needs to go and how carbon dioxide gets shipped off to the lungs for waste removal.

TIP

Remember that the Latin root for blood is *hemo;* other than filtration, which occurs elsewhere, none of the other answer options apply to blood.

3 – 6 Upon inhalation, molecules of **3. oxygen** diffuse into the lung's tissues. From there, these molecules then diffuse into **4. red blood** cells, which contain a pigment called **5. hemoglobin.** Simultaneously, a second substance formed during cellular respiration, **6. carbon dioxide,** is released into the lungs to be expelled during exhalation.

7 The complete deprivation of oxygen to tissues is known as **b. anoxia.** Asphyxia is when the oxygen trouble starts with the lungs, as in the case of injury. Hypoxia can be a result of internal causes that interrupt normal oxygen uptake, and catapoxia is a term we made up to try to throw you off.

8 Anoxia: **d. Without O_2 at the cell level.**

9 Systemic or internal respiration: **c. Gaseous exchange between capillaries and cells.**

10 Asphyxia: **b. O_2 lacking and excessive CO_2.**

11 Hypoxia: **a. Low O_2 content in air breathed in.**

12 Pulmonary, or external, respiration: **e. Gaseous exchange in lungs.**

13 Which of the following statements about the mucous membranes of the nasal cavity is not true? **c. They're composed of stratified squamous epithelium.**

14 Why does the nose have sebaceous glands? **c. Their secretions form a protective layer to warm, moisten, and help filter air.**

15 – 30 Following is how Figure 8-2, the respiratory tract, should be labeled.

15. **c. Nasal cavity;** 16. **e. Nose;** 17. **h. Mouth;** 18. **k. Thyroid cartilage;** 19. **b. Larynx;** 20 **f. Right lung;** 21. **j. Nasopharynx;** 22. **d. Oropharynx;** 23. **g. Epiglottis;** 24. **o. Laryngopharynx;** 25. **a. Esophagus;** 26. **m. Trachea;** 27. **p. Left primary bronchus;** 28. **l. Left lung;** 29. **n. Bronchioles;** 30 **i. Alveoli.**

31 What happens when the arytenoids move? **c. The vocal cords move.** This cartilage is tough to spell and pronounce but easy to move.

32 Epiglottis: **a. Voice box lid**

33 Nasal fossa: **d. Nasal cavity**

34 Medulla oblongata: **b. Respiratory center**

35 C-shaped cartilaginous rings: **c. Prevent collapse of trachea**

36 Thyroid cartilage: **e. Adam's apple**

37 The **d. glottis** is the opening between the two vocal folds and the **d. epiglottis** covers that opening as needed. *Note:* The prefix *epi* means "on, upon, or above."

38 Why can it be painful to fly with a sinus infection? **b. The infection can block the Eustachian tube, causing unequal air pressure on each side of the eardrum.**

39–52 Following is how Figure 8-3, the larynx, should be labeled.

39. **c. Hyoid bone;** 40. **a. Thyroid cartilage;** 41. **g. Laryngeal prominence (Adam's apple);** 42. **d. Epiglottis;** 43. **h. Cuneiform cartilage;** 44. **l. Corniculate cartilage;** 45. **e. Arytenoid cartilage;** 46. **i. Arytenoid muscle;** 47. **b. Cricoid cartilage;** 48. **k. C-shaped tracheal cartilages;** 49. **c. Hyoid bone;** 50. **f. Ventricular fold (false vocal cord);** 51. **a. Thyroid cartilage;** 52. **j. True vocal cord**

53 In which part of the respiratory system is cartilage never found? **c. Bronchioles.** They're so small that they need to be more elastic than cartilage would allow them to be.

54 The capillaries and the **b. alveolar sacs** exchange gases. These sacs are the smallest parts of the lungs, so it makes sense that molecular exchange would take place here.

55–59 The trachea divides into two **55. primary bronchi,** which then divide into **56. secondary bronchi** with a branch going to each lobe of the lung. Upon entering the lobe, each divides into **57. tertiary bronchi,** subdividing into smaller tubes called **58. bronchioles.** They terminate in an elongated sac called the atrium surrounded by **59. alveoli** or air sacs.

60 If a pin were to pierce the body from the outside in the thoracic region, which three structures — in order — would it pass through? **c. Parietal pleura, pleural cavity, visceral pleura.** Note that the question asks you to choose from the lists provided, not from the entire structure of the body.

61–67 Following is how Figure 8-4, the bronchiole, should be labeled.

61. **c. Tertiary bronchiole;** 62. **e. Pulmonary arteriole;** 63. **a. Pulmonary venule;** 64. **f. Alveolar duct;** 65. **b. Alveolar sac;** 66. **g. Alveoli;** 67. **d. Pulmonary capillary**

68 Why do silicosis and anthracosis cause permanent damage to the lungs? **c. Small particles cause scars inside the lungs, and scar tissue isn't as elastic as normal tissue.** Even if you could get all the sand and coal dust out of the lungs, the scarring they caused while embedded there is permanent.

Chapter 9

Fueling the Functions: The Digestive System

*I*t's time to feed your hunger for knowledge about how nutrients fuel the whole package that is the human body. In this chapter, we help you swallow the basics about getting food into the system and digest the details about how nutrients move into the rest of the body. You also get plenty of practice following the nutritional trail from first bite to final elimination.

Digesting the Basics: It's Alimentary!

Before jumping into a discussion of the alimentary tract, we need to review some basic terms:

▶ **Ingestion:** Taking in food

▶ **Digestion:** Changing the composition of food — splitting large molecules into smaller ones — to make it usable by the cells

▶ **Deglutition:** Swallowing, or moving food from the mouth to the stomach

▶ **Absorption:** Occurs when digested food moves through the intestinal wall and into the blood

▶ **Egestion:** Eliminating waste materials or undigested foods at the lower end of the digestive tract; also known as *defecation*

The alimentary tract develops early on in a growing embryo. The primitive gut, or *archenteron,* develops from the *endoderm* (inner germinal layer) during the third week after conception, a stage during which the embryo is known as a *gastrula*. At the anterior end (head end), the oral cavity, nasal passages, and salivary glands develop from a small depression called a *stomodaeum* in the *ectoderm* (outer germinal layer). The anal and urogenital structures develop at the opposite, or posterior, end from a depression in the ectoderm called the *proctodaeum*. In other words, the digestive tract develops from an endodermal tube with ectoderm at each end.

Whereas the respiratory tract (see Chapter 8) is a two-way street — oxygen flows in and carbon dioxide flows out — the digestive tract is designed to have a one-way flow (although when you're sick or your body detects something bad in the food you've eaten, what goes down sometimes comes back up). Under normal conditions, food moves through your body in the following order:

Mouth → Pharynx → Esophagus → Stomach → Small intestine → Large intestine

When you swallow food, it's mixed with digestive enzymes in both saliva and stomach acids. Circular muscles on the inside of the tract and longitudinal muscles along the outside of the tract keep the material moving right through defecation at the end of the line.

Chew on these sample questions about the alimentary tract:

1.–5. Match the alimentary tract terms with their descriptions.

1. _____ Taking in food

2. _____ Elimination of waste

3. _____ Movement of food from mouth to stomach

4. _____ Means of transporting food into the blood

5. _____ Mechanical/chemical changing of food composition

a. Digestion

b. Ingestion

c. Deglutition

d. Absorption

e. Egestion

6.–16. As you read through the rest of this chapter, identify the parts of the digestive system. Referring to Figure 9-1, use the terms that follow to identify the corresponding parts of the digestive system.

a. Pancreas

b. Colon

c. Liver

d. Small intestine

e. Salivary glands

f. Gallbladder

g. Appendix

h. Anus

i. Esophagus

j. Rectum

k. Stomach

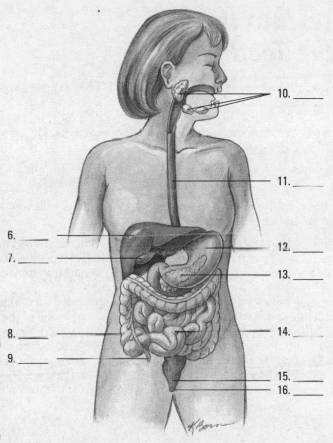

10. _____

11. _____

6. _____

7. _____

12. _____

13. _____

8. _____

14. _____

Figure 9-1: 9. _____
The organs
and glands
of the diges-
tive system.

15. _____
16. _____

Illustration by Kathryn Born, MA

17. What parts of a developing embryo form the alimentary tract?

   a. The placenta

   b. The inner and outer germinal layers

   c. The endoderm and ectoderm layers

   d. Both b and c

18. Identify the correct sequence of the movement of food through the body:

   a. Mouth → Pharynx → Esophagus → Stomach → Small intestine → Large intestine

   b. Mouth → Esophagus → Pharynx → Stomach → Small intestine → Large intestine

   c. Mouth → Pharynx → Esophagus → Stomach → Large intestine → Small intestine

   d. Mouth → Pharynx → Stomach → Esophagus → Small intestine → Large intestine

Nothing to Spit At: Into the Mouth and Past the Teeth

In addition to being very useful for communicating, the mouth serves a number of important roles in the digestive process:

✔ Chewing, formally known as *mastication,* breaks down food mechanically into smaller particles.

✔ The act of chewing increases blood flow to all the mouth's structures and the lower part of the head.

✔ Saliva from *salivary glands* in the mouth helps prepare food to be swallowed and begins the chemical breakdown of carbohydrates.

✔ Taste buds on the tongue stimulate saliva production. Interestingly, studies have shown that taste preferences can change in reaction to the body's specific needs. In addition, the smell of food can get gastric juices flowing in preparation for digestion.

The mouth's anatomy begins, of course, with the lips, which are covered by a thin, modified mucous membrane. That membrane is so thin that you can see the red blood in the underlying capillaries. (That's the unromantic reason for the lips' natural rosy glow.) As you see in the following sections, the mouth itself is divided into two regions defined by the arches of the upper and lower jaws. The *vestibule* is the region between these *dental arches,* cheeks, and lips, whereas the *oral cavity* is the region inside the dental arches. (Figure 9-2 introduces the major structures of the mouth and the pharynx; flip to Chapter 8 for details on the pharynx.)

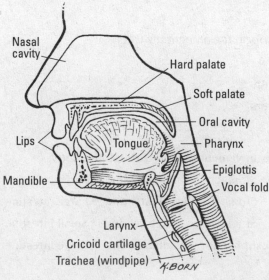

Figure 9-2:
Structures of the mouth and pharynx.

Labels: Nasal cavity, Hard palate, Soft palate, Oral cavity, Lips, Tongue, Pharynx, Mandible, Epiglottis, Vocal fold, Larynx, Cricoid cartilage, Trachea (windpipe)

K.BORN

Illustration by Kathryn Born, MA

Entering the vestibule

The inner surface of the lips is covered by a mucous membrane. Sickle-shaped pieces of tissue called *labial frenula* attach the lips to the gums. Within the mucous membrane are *labial glands,* which produce mucus to prevent friction between the lips and the teeth. The cheeks are made up of *buccinator muscles* and a *buccal pad,* a subcutaneous layer of fat. The buccinator muscles keep the food between the teeth during the act of chewing. Elastic tissue in the mucous membrane keeps the lining of the cheeks from forming folds that would be bitten during chewing (usually — most people have bitten the insides of their cheeks at one time or another).

The dental arches are formed by the *maxillae* (upper jaw) and the *mandible* (lower jaw) along with the *gingivae* (gums) and teeth of both jaws. The gingivae are dense, fibrous tissues attached to the teeth and the underlying jaw bones; they're covered by a mucous membrane extending from the lips and cheeks to form a collar around the neck of each tooth. The gums are very vascular (meaning that lots of blood vessels run through them) but poorly innervated (meaning that, fortunately, they're not generally very sensitive to pain).

Teeth rise from openings in the jawbone called *sockets,* or *alveoli.* You have a number of different kinds of teeth, and each has a specific contribution to the process of biting and chewing. Humans get two sets of teeth in a lifetime. The first temporary, or *deciduous,* set is known as *milk teeth.* Babies between 6 months and 2 years old "cut," or *erupt,* four incisors, two canines, and two molars in each jaw. These teeth are slowly replaced by permanent teeth from about 5 or 6 years of age until the final molars — referred to as *wisdom teeth* — erupt between 17 and 25 years of age.

An adult human has the following 16 teeth in each jaw (for a total set of 32 teeth):

- Four *incisors,* which are chisel-shaped teeth at the front of the jaw for biting into and cutting food

- Two *canines*, or *cuspids,* which are pointed teeth on either side of the incisors for grasping and tearing

- Four *premolars*, or *bicuspids,* which are flatter, shallower teeth that come in pairs just behind the canines

- Six *molars,* which are triplets of broad, flat teeth on either side of the jawbone for grinding and mixing food prior to swallowing

Regardless of type, each tooth has three primary parts:

- **Crown:** The part that projects above the gum

- **Neck:** The region where the gum attaches to the tooth

- **Root:** The internal structure that firmly fixes the tooth in the alveolus (socket)

Teeth primarily consist of yellowish *dentin* with a layer of *enamel* over the crown and a layer of *cementum* over the root and neck, which are connected to the bone by the *periodontal membrane* (called the *periodontal ligament* in some parts of the world). Cementum and dentin are nearly identical in composition to bone; enamel consists of 94 percent calcium phosphate and calcium carbonate and is thickest over the chewing surface of the tooth.

Depending on the structure of the tooth, the root can be a single-, double-, or even triple-pointed structure. In addition, each tooth has a *pulp cavity* at the center that's filled with connective and lymphatic tissue, nerves, and blood vessels that enter the tooth through the root canal via an opening at the apex of the root of the tooth called the *apical foramen.* Now you know why it hurts so much when dentists have to drill down and take out that part of an infected tooth!

Moving along the oral cavity

The roof of the oral cavity is formed by both the *hard palate,* a bony structure covered by fibrous tissue and the ever-present mucous membrane, and the *soft palate,* a movable partition of fibromuscular tissue that prevents food and liquid from getting into the nasal cavity. (It's also the tissue that sometimes vibrates in sleep, causing a sonorous grating sound referred to as *snoring.*) The soft palate hangs at the back of the oral cavity in two curved folds that form the palatine arches. The *uvula,* a soft conical process (or piece of tissue), hangs in the center between those folds. (You can see the hard and soft palates in Figure 9-2.)

Beyond the soft palate, the *palatopharyngeal* (or *pharyngopalatine*) arch curves sharply toward the midline and blends with the wall of the pharynx, ending at the *dorsum* (back) of the tongue. Another structure, the anterior *palatoglossal* (or *glossopalatine*) arch, starts on the surface of the palate at the base of the uvula and continues in a wide curve forward and downward, ending next to the posterior (back) one-third of the tongue. At the base of these arches and between the folds lie the *palatine tonsils* — if a surgeon hasn't removed them because of frequent childhood infections. The *faucial isthmus* or *oropharynx* is the junction between the oral cavity and the pharynx. The oropharynx opens during swallowing and closes when you move the dorsum of the tongue against the soft palate when breathing.

The following structures detail some major parts of the oral cavity: the tongue and the salivary glands.

The tongue

As shown in Figure 9-2, the tongue, which is a tight bundle of interlaced muscles, and its associated mucous membrane form the floor of the oral cavity. Two distinct groups of muscles — *extrinsic* and *intrinsic* — are used in tandem for mastication (chewing), deglutition (swallowing), and to articulate speech.

- **The extrinsic muscles:** Used to move the tongue in different directions, the extrinsic muscles originate outside the tongue and are attached to the mandible, styloid processes of the temporal bone and the hyoid and, along with a fold of mucous membrane called the *lingual frenulum,* anchor the tongue.

- **The intrinsic muscles:** A complex muscle network, the intrinsic muscles allow the tongue to change shape for talking, chewing, and swallowing.

Three primary types of *papillae* (nipple-shaped protrusions) cover the tongue's forward upper surface:

- **Filiform papillae** are fine, brushlike papillae that cover the dorsum, the tip, and the lateral margins of the tongue. They're the most numerous papillae and don't hold any taste buds.

✓ **Fungiform papillae** are large, red, mushroom-shaped papillae scattered among the filiform papillae. They have taste buds, which are special receptors that communicate taste signals to the brain.

✓ **Vallate papillae,** also called *circumvallate papillae,* are flattened structures, each with a moatlike trough ringing it and containing taste buds down the side. There are 12 of these on the tongue, and they surround a V-shaped furrow toward the back of the tongue called the *sulcus terminalis.* The furrow divides the anterior from the posterior part of the tongue.

There are no papillae on the back (posterior) one-third of the tongue; that part has only a mucous membrane covering lymphatic tissue, which forms the *lingual tonsils.*

The salivary glands

The oral cavity has three pairs of salivary (exocrine) glands producing saliva:

✓ The *parotid* salivary glands are the largest salivary glands and are found in the posterior region of the mandible, just in front of and below each ear. They secrete a serous fluid through the parotid duct (which is opposite the second upper molar tooth) that makes up about 25 percent of the saliva that enters the oral cavity.

✓ The smallest pair of the trio, the *sublingual* salivary glands, lies near the tongue under the oral cavity's mucous membrane floor to release secretions through eight to twenty excretory ducts directly onto the mucous membrane. These produce about 5 percent of the saliva that enters the oral cavity.

✓ The *submandibular* (or *submaxillary*) salivary glands are posterior to the sublingual, along the side of the lower jaw. They're about the size of walnuts and release fluid onto the floor of the mouth, under the tongue. They produce approximately 70 percent of the saliva that enters the oral cavity.

And those secretions are nothing to spit at. Saliva does the following:

✓ Dissolves and lubricates food to make it easier to swallow

✓ Contains *ptyalin,* or *salivary amylase,* an enzyme that initiates chemical digestion of certain carbohydrates

✓ Contains *lysozymes* that break down the carbohydrate in the cell wall of bacteria, killing the bacteria

✓ Moistens and lubricates the mouth and lips, keeping them pliable and resilient for speech and chewing

✓ Frees the mouth and teeth of food, foreign particles, and epithelial cells

✓ Produces the sensation of thirst to prevent you from becoming dehydrated

Following are some practice questions regarding the vestibule and oral cavity:

Q. The function of the mouth is

    a. to mix solid foods with saliva.

    b. to break down milk protein via the enzyme rennin.

    c. To masticate or break down food into small particles.

    d. a and c

    e. a, b, and c

A. The correct answer is to mix solid foods with saliva and mastication (a and c). The mouth does lots of things, including mixing saliva into the food to add the enzyme ptyalin, but that's not rennin. With answer options like these, it's best to stick to the basics.

19.–30. Use the terms that follow to identify the parts of a tooth, as shown in Figure 9-3.

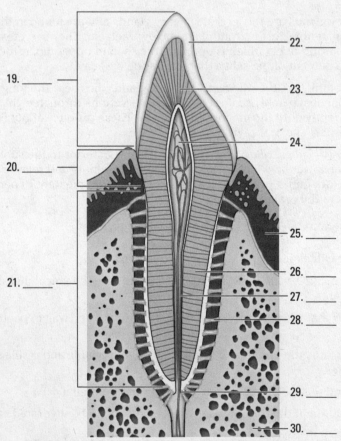

19. _____
20. _____
21. _____
22. _____
23. _____
24. _____
25. _____
26. _____
27. _____
28. _____
29. _____
30. _____

Figure 9-3: The composition of a tooth.

Illustration by Imagineering Media Services Inc.

    a. Root canal

    b. Neck

    c. Bone

    d. Dentin

    e. Crown

> **f.** Periodontal ligament
>
> **g.** Gingiva
>
> **h.** Enamel
>
> **i.** Root
>
> **j.** Apical foramen
>
> **k.** Pulp cavity
>
> **l.** Cementum

31. What's the function of the soft palate, which vibrates when you snore?

> **a.** It separates the uvula from the periodontal ligament.
>
> **b.** It prevents premature swallowing of food before it's thoroughly masticated.
>
> **c.** It forces saliva back over the tongue.
>
> **d.** It prevents food and liquid from getting into the nasal cavity.

32. What anchors the tongue?

> **a.** The hard and soft palates
>
> **b.** The intrinsic muscles and the sulcus terminalis
>
> **c.** The extrinsic muscles and the lingual frenulum
>
> **d.** The mandibular arches

33. Which of the following statements about teeth is *not* true?

> **a.** The permanent teeth in each human jaw are four incisors, two canines, four premolars, and six molars.
>
> **b.** Each tooth has a single cuspid anchoring it.
>
> **c.** The tooth cavity contains the tooth pulp.
>
> **d.** The enamel consists of 94 percent calcium phosphate and calcium carbonate.

34.–38. Match each description with the proper anatomical structure.

> **34.** _____ Soft conical process projecting from the soft palate
>
> **35.** _____ The junction between the mouth and pharynx
>
> **36.** _____ Forms a collar around the teeth and is poorly innervated
>
> **37.** _____ Sharply curved arch that bends laterally with the walls of the pharynx
>
> **38.** _____ Arch that starts at the buccal surface of the palate at the base of the uvula and ends alongside the back third of the tongue

> **a.** Pharyngopalatine arch
>
> **b.** Faucial isthmus
>
> **c.** Gingivae
>
> **d.** Glossopalatine arch
>
> **e.** Uvula

39. If it hasn't been surgically removed, where can the palatine tonsil be found?

   a. In the posterior wall of the pharynx

   b. In the smooth posterior third of the tongue

   c. In the region between the rigid hard palate and the soft palate

   d. In the region between the palatopharyngeal and palatoglossal arches

40. Besides making it possible to spit, why do people have saliva?

   a. To facilitate swallowing

   b. To initiate the digestion of certain carbohydrates

   c. To moisten and lubricate the mouth and lips

   d. All of the above

41. What role does saliva play in preventing infection?

   a. The moisture overwhelms viruses, killing them.

   b. Its lysozymes break down cell walls, killing bacteria.

   c. Ptyalin absorbs nutrients from microbes, rendering them inert.

   d. Salivary amylase blocks bacteria's ability to multiply.

42.–44. Match the descriptions with the anatomical structures.

   42. _____ Fine, brushlike structures found covering the dorsum of the tongue

   43. _____ Large, mushroom-shaped structures

   44. _____ Large structures, each surrounded by a moat, that form a V-shaped furrow in the tongue

   a. Vallate papillae

   b. Filiform papillae

   c. Fungiform papillae

Stomaching the Body's Fuel

Deglutition (swallowing) occurs in three phases:

1. **The tip of the tongue elevates slightly, pushing against the hard palate, sliding food onto the back of the tongue, and ultimately propelling it toward the pharynx.**

   You can see these parts of the body in Figure 9-2.

2. ***Tensor muscles* tighten the palate while *levator muscles* raise it until the palate meets the pharyngeal wall, sealing off the nasopharynx from the oropharynx.**

   This action momentarily stops breathing and ensures that food and fluid don't regurgitate through the nose — unless someone makes you laugh, of course.

3. **The *bolus* (food mass) heads "down the hatch."**

   The "hatch" is borrowed nautical slang for the *esophagus*.

The pharynx is an oval, fibrous, muscular sac, about 5 inches long. It opens into the nasal cavity, the oral cavity, the larynx, and the esophagus. On the lateral walls of the nasopharynx are located the openings to the Eustachian tubes, which connect with the middle ear. In the posterior wall is a mass of lymphatic tissue, the pharyngeal tonsil or adenoid.

The *esophagus,* which is approximately 10 inches long and ½ inch in diameter, carries food through three body regions: the neck, the thorax, and the abdomen. It's not a straight tube, but rather curves slightly to the left as it passes through the diaphragm 1 inch to the left of the midline. The very thick walls of the esophagus are lined with non-keratinized, stratified squamous epithelium and include a fibrous outer layer made up of elastic fibers that permit *distention* during swallowing. Most of the esophageal muscle is smooth, although striated muscle — which is subject to voluntary control — predominates in the upper third. There are two sphincters, or muscular rings. The upper gastroesophageal sphincter is under conscious control and used during breathing, eating, belching, and vomiting. It prevents food from going down the windpipe. A muscular layer contains both longitudinal and circular layers of smooth muscle. The circular layers contract in sequence, like a series of shrinking and expanding rings, in a movement called *peristalsis* that forces the bolus downward. The longitudinal layers act in concert with the circular muscles, pulling the esophagus over the bolus as it moves downward.

All this pushing and pulling ultimately releases the bolus into the stomach, a pear-shaped bag of an organ that lies just beneath the ribs and diaphragm. About 1 foot long and ½ foot wide, a human stomach's normal capacity is about 1 quart. When empty, the mucous lining lies in folds called *rugae;* rugae allow expansion of the stomach when you gorge and shrink when it's empty to decrease the surface area exposed to acid. Food enters the upper end of the stomach, called the *cardiac* region, through a ring of muscles called the *cardiac sphincter,* or the lower *gastroesophageal sphincter,* which generally remains closed to prevent gastric acids from moving up into the esophagus. (When gastric acids do move into places they don't belong, the painful sensation is referred to as "heartburn," which may help you remember the term "cardiac sphincter.") The dome-shaped area below the cardiac region is called the *fundus region;* it expands superiorly with really big meals. The lower part of the stomach, shaped like the letter J, is the *pylorus.* The middle part, or *body,* of the stomach forms a large curve on the left side called the *greater curvature.* The right, much shorter, border of the stomach's body is called the *lesser curvature.* The far end of the stomach remains closed off by the *pyloric sphincter* until its contents have been digested sufficiently to pass into the *duodenum* of the small intestine.

The wall of the stomach consists of three layers of smooth muscle lined by mucous membrane and covered by the peritoneum, a serous membrane (also known as serosa). The fibers of the outer layer of muscle run longitudinally, the middle layer of muscle consists of circular fibers that encircle the stomach, and the inner layer of muscle fibers runs obliquely only along the fundus region. The stomach's mucous membrane is covered with simple columnar epithelium (see Chapter 4 for more about this tissue) containing mucous glands, protecting the *gastric mucosa* from the acid and enzymes.

The three types of gastric glands in the stomach's *epithelium* (lining) are

- **Cardiac glands:** Found in the cardiac region (of course)
- **Pyloric glands:** Secrete mucous in the pyloric region
- **Fundic glands:** Lined with chief cells and parietal cells and are located throughout the stomach's body and fundus

The three types of cells in the *mucosa* (lining) of the stomach are

- **Mucous cells:** Secrete *mucin* (mucous) to protect the mucosa from the high acidity of the gastric juices.

- **Chief cells:** Secrete *pepsinogen,* a precursor to the enzyme pepsin that helps break down certain proteins into peptides. (Chief cells in children also produce an enzyme called *rennin,* not found in adults, which acts upon milk proteins.)

- **Parietal cells:** Lie alongside chief cells and secrete the hydrochloric acid, or HCl, that combines with pepsinogen to form pepsin to catalyze protein digestion.

The peristaltic contractions that get the bolus into the stomach aren't limited to the esophagus. Instead, peristalsis continues into the musculature of the stomach and stimulates the release of a hormone called *gastrin.* Within minutes, gastrin triggers secretion of gastric juices that reduce the bolus of food to a thick semiliquid mass called *chyme,* which passes through the pyloric sphincter into the small intestine within one to four hours of the food's consumption.

Gastric juices are thin, colorless fluids with an extremely acid pH that ranges from 1 to 4 (see Chapter 1 for details on pH). The quantity of acid released depends on the amount and type of food being digested.

One more part attached to the stomach that we should mention is the *greater omentum.* This is a peritoneal fold that hangs like an apron from the greater curvature of the stomach all the way down to the transverse colon, covering all the small intestine and most of the large intestine. Also called a *caul* or *velum,* this lining can be laden with fat, particularly in obese people.

45.–52. Use the terms that follow to identify the anatomy of the stomach, as shown in Figure 9-4.

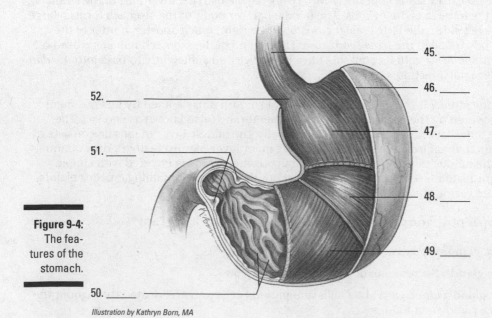

45. _____
46. _____
47. _____
48. _____
49. _____
50. _____
51. _____
52. _____

Figure 9-4: The features of the stomach.

Illustration by Kathryn Born, MA

    a. Circular muscle layer

    b. Esophagus

    c. Rugae of the mucosa

    d. Cardiac (lower gastroesophageal) sphincter

    e. Serosa

    f. Oblique muscle layer

    g. Pyloric sphincter

    h. Longitudinal muscle layer

53. What does peristalsis do to food?

    a. These contractions of the small intestine wring nutrients from partially digested food.

    b. This reflexive upward muscular motion mixes food with stomach acids to create vomit.

    c. This grinding action of cartilaginous tissue in the esophagus breaks down food before it enters the stomach.

    d. This sequential contraction of circular muscles helps move food through the esophagus.

54. Which digestive tract sphincter is under voluntary control?

    a. The fundic sphincter

    b. The upper gastroesophageal sphincter

    c. The pyloric sphincter

    d. The ileocolic sphincter

55. Which part of the digestive tract isn't doing its job when you have heartburn?

    a. The esophagus

    b. The pylorus

    c. The peritoneal fold

    d. The cardiac sphincter

56. Where can chyme first be found along the digestive tract?

    a. The pylorus

    b. The peritoneum

    c. The lower gastroesophageal sphincter

    d. The large intestine

Breaking Down the Work of Digestive Enzymes

So what exactly does all the work of digesting and breaking down food? That question brings you back into the realm of proteins (which we introduce in Chapter 1). Proteins that are *enzymes* act as catalysts, meaning that they initiate and accelerate chemical reactions without themselves being permanently changed in the reaction. Enzymes are very picky proteins indeed; they are effective only in their own pH range, they catalyze only a single chemical reaction, they act on a specific substance called a *substrate,* and they function best at 98.6 degrees Fahrenheit, which just happens to be normal body temperature.

The following sections take you on a tour of the organs that produce digestive enzymes.

The small intestine

Most enzyme reactions — in fact most digestion and practically all absorption of nutrients — takes place in the small intestine. Stretching 7 meters (which is nearly 23 feet!), this long organ extends from the stomach's pylorus to the *ileocecal junction* (the point where the small intestine meets the large intestine), gradually diminishing in diameter along the way.

Three regions of the small intestine play unique roles as chyme moves through them:

- ✔ **Duodenum:** The first section of the small intestine is also the shortest and widest section. As partially digested food enters the duodenum, its acidity stimulates the intestine to secrete the intestinal hormone *enterocrinin,* which controls the secretion of intestinal juices, stimulates the pancreas to secrete its juices, and stimulates the liver to secrete bile. Both the liver and pancreas share a common opening into the duodenum. Lined with large and numerous *villi* (fingerlike projections), the duodenum also has *Brunner's glands* that secrete a clear, alkaline mucus. The glands are most numerous near the entry to the stomach and decrease in number toward the opposite, or *jejunum,* end.

- ✔ **Jejunum:** This is where the bulk of the absorption of nutrients occurs. This region of the small intestine also contains villi, but unlike the duodenum, it has numerous large circular folds at the beginning that decrease in number toward the *ileum* end.

- ✔ **Ileum:** Any nutrients not absorbed in the jejunum are absorbed in the ileum before the food passes into the large intestine. Another feature within the ileum are *Peyer's patches,* which are aggregates of lymph nodules that line this region of the small intestine, becoming largest and most numerous at the distal end. They monitor the gut for pathogenic microorganisms to facilitate the mucosa's generation of immune responses. The ileum opens into the *cecum* of the large intestine through the *ileocaecal valve.*

A microscopic look at the small intestine reveals circular folds called *plicae circularis,* which project 3 to 10 millimeters into the intestinal lumen, or cavity, and extend anywhere from halfway to entirely around the tube. These are permanent folds that don't smooth when the intestine is distended. Also present are villi that greatly increase the surface area through which the small intestine can absorb nutrients. Each villus contains a network of capillaries and a central lymph vessel, or *lacteal,* which absorbs fatty acids, to form a milk-white substance called *chyle.* The chyle is carried through lymph vessels to the thoracic duct, which empties into the subclavian vein. Simple sugars, amino acids, vitamins, minerals, and water are absorbed by the villi through the mucosa. The surface of the villus is simple columnar epithelium (if you can't recall what that means, flip to Chapter 4). Electron microscopy, which can magnify tissues far more than an optical microscope can, reveals that the surface of each villus is further increased by *microvilli.* Peristalsis continues into the small intestine, shortening and lengthening the villi to mix intestinal juices with food and increase absorption. Intestinal glands lie in the depressions between villi, and packed inside these glands are antimicrobial *Paneth cells* within glands called the *crypts of Lieberkühn,* which secrete enzymes that assist pancreatic enzymes.

Intestinal juices contain three types of enzymes:

- ✔ **Enterokinase** has no enzyme action by itself, but when added to pancreatic juices, it combines with *trypsinogen* to form *trypsin,* which can break down proteins.

- ✔ **Erepsins, or proteolytic enzymes,** don't directly digest proteins but instead complete protein digestion started elsewhere. They split polypeptide bonds, separating amino acids.

- ✔ **Inverting enzymes** split disaccharides into monosaccharides as follows:

Enzyme	*Disaccharide*	*Monosaccharides*
Maltase	Maltose	Glucose + Glucose
Lactase	Lactose	Glucose + Galactose
Sucrase	Sucrose	Glucose + Fructose

The liver

The largest gland in the body, the liver (shown in Figure 9-5 and also well-marked in Figure 10-7 in Chapter 10) is divided into a large right lobe and a small left lobe by the *falciform ligament,* another peritoneal fold. Two smaller lobes — the *quadrate* and *caudate* lobes — are found on the lower (inferior) and back (posterior) sides of the right lobe. The quadrate lobe surrounds and cushions the *gallbladder,* a pear-shaped structure that stores and concentrates *bile,* which it empties periodically through the *cystic duct* to the *common bile duct* and on into the duodenum during digestion. Bile aids in the digestion and absorption of fats; it consists of bile pigments, bile salts, and cholesterol.

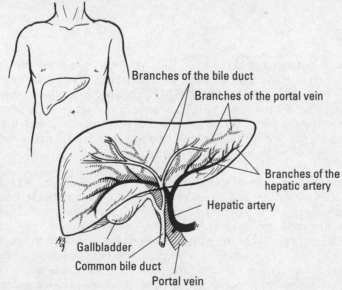

Figure 9-5: A closer look at the liver.

Branches of the bile duct

Branches of the portal vein

Branches of the hepatic artery

Hepatic artery

Gallbladder

Common bile duct

Portal vein

Illustration by Kathryn Born, MA

The liver secretes diluted bile through the right and left *hepatic ducts* into the common hepatic duct that joins the cystic duct coming from the gallbladder, forming the common bile duct. Liver tissue is made up of rows of cuboidal cells (see Chapter 4) separated by microscopic blood spaces called *sinusoids.* Blood from the *interlobular veins and arteries* circulates through the sinusoids with food and oxygen for the liver cells, picking up materials along the way. The blood then enters the *intralobular veins,* which carry it to the *sublobular veins,* which empty into the *hepatic vein,* which leads to the *inferior vena cava.* Bile secreted from the liver cells is carried by *biliary canaliculi* (bile capillaries) to the bile ducts and then to the hepatic ducts, eventually emptying into the duodenum.

Considering the number of vital roles the liver plays, the complexity of that process isn't too surprising. Among the liver's various functions are

- Production of blood plasma proteins including *albumin,* antibodies to fend off disease, a blood anticoagulant called *heparin* that prevents clotting, and bile pigments from red blood cells, the yellow pigment *bilirubin,* and the green bile pigment *biliverdin*

- Storage of vitamins and minerals as well as glucose in the form of glycogen

- Conversion and utilization through enzyme activity of fats, carbohydrates, and proteins

- Filtering and removal of nonfunctioning red blood cells, toxins (isolated by *Kupffer cells* in the liver), and waste products from amino acid breakdown, such as urea and ammonia

Unfortunately, a number of serious diseases can damage the liver. The hepatitis virus inflames the gland, and cirrhosis caused by repeated toxic injury (often through alcohol or other substance abuse) destroys Kupffer cells and replaces them with scar tissue. Also, painful gallstones can develop when cholesterol clumps together to form a center around which the gallstone can form.

The pancreas

Equally important, though not as large as the liver, the *pancreas* looks like a roughly 7-inch long, irregularly shaped prism. It has a broad head lodged in the curve of the duodenum. The head is attached to the body of the gland by a slight constriction called the neck, and the opposite end gradually tapers to form a tail. The pancreatic duct extends from the head to the tail, receiving the ducts of various lobules that make up the gland. It generally joins the common bile duct, but some 40 percent of humans have a pancreatic duct and a common bile duct that open separately into the duodenum.

Uniquely, the pancreas is both an *exocrine gland,* meaning that it releases its secretion externally either directly or through a duct, and an *endocrine gland,* meaning that it produces hormonal secretions that pass directly into the bloodstream without using a duct. However, most of the pancreas is devoted to being an exocrine gland secreting pancreatic juices into the duodenum. The endocrine portion of the gland secretes insulin vital to the control of sugar metabolism in the body through small, scattered clumps of cells known as *islets of Langerhans.* Because it contains sodium bicarbonate, pancreatic juice is alkaline, or base, with a pH of 8. Enzymes released by the pancreas act upon all types of foods, making its secretions the most important to digestion. Its enzymes include pancreatic amylase, or carbohydrate enzymes; pancreatic lipase, or fat enzymes; trypsin, or protein enzymes; and nuclease, or nucleic acid enzymes.

The most commonly known pancreatic disease is called diabetes mellitus, or sugar diabetes, which occurs when the islets of Langerhans cease producing insulin. Without insulin, the body can't use sugar, which builds up in the blood and is excreted by the kidneys.

The large intestine

After chyme works its way through the small intestine, it then must move through 5 feet or so of large intestine. The byproduct of the small intestine's work enters at the *ileocaecal valve* and then moves through the following regions of the large intestine:

Cecum → Ascending colon→ Transverse colon → Descending colon → Sigmoid colon → Rectum → Anus

The large intestine is about 3 inches wide at the start and decreases in width all the way to the anus. As the unabsorbed material moves through the large intestine, excess water is reabsorbed, drying out the material. In fact, most of the body's water absorption takes place in the large intestine. Peristaltic movement continues, albeit rather feebly, in the cecum and ascending colon. The large intestine has a longitudinal muscle layer in the form of three bands running from the cecum to the rectum called the *taenia coli.* The large intestine serves no digestive function and secretes only mucus. It has no villi, nor does it have any intestinal glands. Truly, it is the end of the line.

That's a lot of material to digest. See how much you remember:

57. Which of the following terms doesn't belong?

a. Enterokinase

b. Maltose

c. Amylase

d. Sucrase

58. As the largest gland in the body, the liver serves a number of vital roles. This is *not* one of them:

a. Production of insulin

b. Production of bile pigments

c. Storage of vitamins and minerals

d. Removal of old blood cells

59. Where are most nutrients absorbed?

a. The pylorus

b. The jejunum

c. The ileocecus

d. The fundus

60. What role do villi play throughout the small intestine?

   a. They neutralize acids.

   b. They move chyme along the intestinal route.

   c. They increase the surface area available to absorb nutrients.

   d. They further masticate the bolus.

61. What is the role of enterokinase in digestion?

   a. It activates a pancreatic enzyme that can break down protein.

   b. It acts as a proteolytic enzyme.

   c. It prompts the release of erepsin.

   d. It inverts the functions of other enzymes.

62. Why are pancreatic secretions the most important in digestion?

   a. They neutralize the acids coming from the stomach.

   b. They contain enzymes that break down all types of foods.

   c. They form the plicae circularis.

   d. They invert the functions of other digestive secretions.

63. True or false: The large intestine performs a role in digestion.

Answers to Questions on the Digestive Tract

The following are answers to the practice questions presented in this chapter.

1 Taking in food: **b. Ingestion.**

2 Elimination of waste: **e. Egestion.**

3 Movement of food from mouth to stomach: **c. Deglutition.**

4 Means of transporting food into the blood: **d. Absorption.**

5 Mechanical/chemical changing of food composition: **a. Digestion.**

6–16 Following is how Figure 9-1, the digestive system, should be labeled.

6. **c. Liver;** 7. **f. Gallbladder;** 8. **b. Colon;** 9. **g. Appendix;** 10. **e. Salivary glands;**
11. **i. Esophagus;** 12. **k. Stomach;** 13. **a. Pancreas;** 14. **d. Small intestine;** 15. **j. Rectum;**
16. **h. Anus**

17 What parts of a developing embryo form the alimentary tract? **d. Both b and c.** The tube that becomes the digestive tract develops from the endoderm with ectoderm at each end; both of these parts are also known as germinal layers.

18 Identify the correct sequence of the movement of food through the body: **a. Mouth → Pharynx → Esophagus → Stomach → Small intestine → Large intestine.**

Although remembering the sequence M-P-E-S-small-large can be helpful, you can also try this phrase to jog your memory: Most Phones Enable Speeches, from Small to Large.

19–30 Following is how Figure 9-2, the tooth, should be labeled.

19. **e. Crown;** 20. **b. Neck;** 21. **i. Root;** 22. **h. Enamel;** 23. **d. Dentin;** 24. **k. Pulp cavity;**
25. **g. Gingiva;** 26. **l. Cementum;** 27. **a. Root canal;** 28. **f. Periodontal ligament;** 29. **j. Apical foramen;** 30. **c. Bone**

31 What's the function of the soft palate, which vibrates when you snore? **d. It prevents food and liquid from getting into the nasal cavity.** So, as irritating as snoring may be, you need that soft palate to stay put.

32 What anchors the tongue? **c. The extrinsic muscles and the lingual frenulum.** Extrinsic means "outside," and anchor points occur outside, not inside (or intrinsically).

33 Which of the following statements about teeth is *not* true? **b. Each tooth has a single cuspid anchoring it.** You can rule out this answer option as false because a cuspid is a type of tooth, so it makes no sense that each tooth would have another type of tooth anchoring it.

34 Soft conical process projecting from the soft palate: **e. Uvula.**

35 The junction between the mouth and pharynx: **b. Faucial isthmus.**

36 Forms a collar around the teeth and is poorly innervated: **c. Gingivae.**

37 Sharply curved arch that bends laterally with the walls of the pharynx: **a. Pharyngopalatine arch.**

38 Arch that starts at the buccal surface of the palate at the base of the uvula and ends alongside the back third of the tongue: **d. Glossopalatine arch.**

39 If it hasn't been surgically removed, where can the palatine tonsil be found? **d. In the region between the palatopharyngeal and palatoglossal arches.**

40 Besides making it possible to spit, why do people have saliva? **d. All of the above.** Multifunctional stuff, that saliva. It facilitates swallowing, initiates the digestion of certain carbohydrates, and moistens and lubricates the mouth and lips.

41 What role does saliva play in preventing infection? **b. Its lysozymes break down cell walls, killing bacteria.** Don't forget that bacteria have cell walls, while human cells do not.

42 Fine, brushlike structures found covering the dorsum of the tongue: **b. Filiform papillae.**

43 Large, mushroom-shaped structures: **c. Fungiform papillae.**

44 Large structures, each surrounded by a moat, that form a V-shaped furrow in the tongue: **a. Vallate papillae.**

45–52 Following is how Figure 9-3, the stomach, should be labeled.

45. **b. Esophagus;** 46. **e. Serosa;** 47. **h. Longitudinal muscle layer;** 48. **a. Circular muscle layer;** 49. **f. Oblique muscle layer;** 50. **c. Rugae of the mucosa;** 51. **g. Pyloric sphincter;** 52. **d. Cardiac (lower gastroesophageal) sphincter.**

53 What does peristalsis do to food? **d. This sequential contraction of circular muscles helps move food through the esophagus.**

A bit of Greek may help you remember this term, which comes from the word *peristaltikos,* which means "to wrap around."

54 Which digestive tract sphincter is under voluntary control? **b. The upper gastroesophageal sphincter.**

55 Which part of the digestive tract isn't doing its job when you have heartburn? **d. The cardiac sphincter.** Biggest clue? Cardiac = heart.

56 Where can chyme first be found along the digestive tract? **a. The pylorus.** Chyme is the thick, semiliquid mass of food that's ready to leave the stomach, so you can select the right answer if you remember your Greek prefixes and suffixes: *pyl–* means "gate," and *–orus* means "guard."

Here's a silly but effective memory tool for this term: When food is ready to leave the stomach, it rings a chime.

57 Which of the following terms doesn't belong? **b. Maltose.** This is a sugar, whereas the other answer options are all enzymes.

58 As the largest gland in the body, the liver serves a number of vital roles. This is *not* one of them: **a. Production of insulin.** That's what the pancreas does.

59 Where are most nutrients absorbed? **b. The jejunum.** If you were looking for the answer "small intestine," you were on the right track. The jejunum is the only part of the small intestine included in the list of answers.

To remember this one, keep in mind that each of the three sections of the small intestine plays a unique role. The duodenum neutralizes the stomach acid as the chyme enters the intestine (among other things), the jejunum absorbs most of the nutrients, and the ileum monitors for pathogens (or keeps you from getting "ill" — "il"eum. Get it?).

60 What role do villi play throughout the small intestine? **c. They increase the surface area available to absorb nutrients.** They do other things along the way, but none of those functions are included in the other answers.

61 What is the role of enterokinase in digestion? **a. It activates a pancreatic enzyme that can break down protein.**

It's tricky to remember which of these enzymes is inactive until it combines with something else. You can either try to memorize the function of each enzyme, or you can pick apart the terms. The prefix *entero–* comes from the Greek word for "intestine." The suffix *–kinase* stems from the Greek word for "moving." "Moving through the intestine" sounds like a good guess, don't you think?

62 Why are pancreatic secretions the most important in digestion? **b. They contain enzymes that break down all types of foods.** Those across-the-board functions make them most critical.

63 True or false: The large intestine performs a role in digestion. **False.** Every nutrient you're going to get out of the food you eat already has been absorbed by the time it reaches the large intestine. It's the body's primary location for reabsorbing water, but that's not a digestive role.

Chapter 10

Spreading the Love: The Circulatory System

In This Chapter

▶ Understanding the heart's rhythm and structure

▶ Identifying the heart's chambers

▶ Checking out the heart's conduction system

▶ Tracing arteries, veins, and capillaries

This chapter gets to the heart of the well-oiled human machine to see how its central pump is the hardest-working muscle in the entire body. From a month after you're conceived to the moment of your death, this phenomenal powerhouse pushes a liquid connective tissue — blood — and its precious cargo of oxygen and nutrients to every nook and cranny of the body, and then it keeps things moving to bring carbon dioxide and waste products back out again. In the first seven decades of human life, the heart beats roughly 2.5 billion times. Do the math: How many pulses has your ticker clocked if the average heart keeps up a pace of 72 beats per minute, 100,000 per day, or roughly 36 million per year?

Moving to the Beat of a Pump

Also called the *cardiovascular system,* the circulatory system includes the heart, all blood vessels, and the blood that moves endlessly through it all (see Figure 10-1). It's referred to as a *closed double system.* The term "closed" is used for three reasons:

✔ The blood is contained in the heart and its vessels.

✔ The vessels specifically target the blood to the tissues.

✔ The heart critically regulates blood flow to the tissues.

The system is called "double" because there are two distinct circuits and cavities within the heart separated by a wall of muscle called the *septum.* (Each cavity in turn has two chambers called *atria* on top and *ventricles* below, for a total of four chambers.) The double circuit moves blood through the heart twice during each circulation cycle, separating oxygenated and deoxygenated blood. Doubling up also compensates for the loss of pressure and flow rate that occurs in the lungs' capillaries, and it pumps up the pressure to the body's extremities. The double circuits are the following:

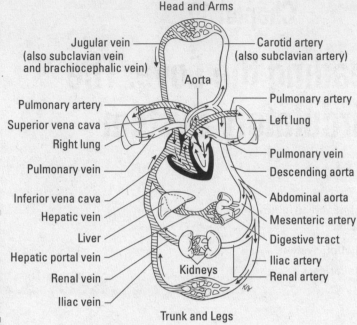

Head and Arms

Jugular vein
(also subclavian vein
and brachiocephalic vein)

Carotid artery
(also subclavian artery)

Aorta

Pulmonary artery

Superior vena cava

Right lung

Pulmonary vein

Inferior vena cava

Hepatic vein

Liver

Hepatic portal vein

Renal vein

Iliac vein

Pulmonary artery

Left lung

Pulmonary vein

Descending aorta

Abdominal aorta

Mesenteric artery

Digestive tract

Iliac artery

Renal artery

Kidneys

Trunk and Legs

Figure 10-1:
The circulatory system is a closed double system.

Illustration by Kathryn Born, MA

- ✔ **Pulmonary:** The *pulmonary circuit* carries blood to and from the lungs for gaseous exchange. Centered in the right side of the heart, this circuit receives deoxygenated blood saturated with carbon dioxide from the veins and pumps it through the *pulmonary trunk* to *pulmonary arteries* to capillaries in the lungs, where the carbon dioxide departs the system. That same blood, freshly loaded with oxygen *(oxygenated)* but with decreased pressure, then returns to the left side of the heart through the *pulmonary veins* where it enters the second circuit. ***Note:*** This is the only case where an artery carries carbon dioxide and a vein carries oxygen.

- ✔ **Systemic:** The *systemic circuit* uses the oxygenated blood to maintain a constant internal pressure and flow rate around the body's tissues. From the left side of the heart, the blood moves with increased pressure and flow rate through the *aorta* first to the coronary arteries and then to a variety of systemic arteries for distribution throughout the body. After oxygen diffuses out of the tissue capillaries to the cells of the body and carbon dioxide diffuses in, the deoxygenated blood returns to the pulmonary circuit on the right side of the heart via the *superior* and *inferior venae cavae* (the singular is *vena cava*).

Although cutely depicted in popular culture as uniformly curvaceous, the heart actually looks more like a blunt, muscular cone (roughly the size of a fist) resting on the diaphragm. A fluid-filled, fibrous sac called the *pericardium* (or *heart sac*) wraps loosely around the package; it's attached to the large blood vessels emerging from the heart. The sternum (breastbone) and third to sixth costal cartilages of the ribs provide protection in front of (ventrally to) the heart. Behind it lie the fifth to eighth thoracic vertebrae. (Check out more about the skeleton in Chapter 5.) Two-thirds of the heart lies to the left of the body's center, with its *apex* (cone) pointed down and to the left.

The biggest part that's farthest away from the apex is called the *base*. The base contains the two atria and the great vessels (aorta and pulmonary trunk) and their semilunar ("half-moon") valves that prevent backflow. At less than 5 inches long and a bit more than 3 inches wide, an adult human heart weighs around 10 ounces — a couple ounces shy of a can of soda.

Three layers make up the wall of the heart:

✔ **Epicardium:** On the outside lies the *epicardium* (or *visceral pericardium*), which is composed of fibroelastic connective tissue containing adipose tissue (fat) that fills external grooves called *sulci* (the singular is *sulcus*). The larger coronary vessels and nerves are found in the adipose tissue that fills the sulci.

✔ **Myocardium:** Beneath the epicardium lies the *myocardium,* which is composed of layers and bundles of cardiac muscle tissue.

✔ **Endocardium:** The *endocardium,* the heart's interior lining, is composed of simple squamous endothelial cells.

Too much to remember? To keep the layers straight, turn to the Greek roots. *Epi–* is the Greek term for "upon" or "on" whereas *endo–* comes from the Greek *endon* meaning "within." The medical definition of *myo–* is "muscle." And *peri–* comes from the Greek term for "around" or "surround." Hence the *epi*cardium is *on* the heart, the *endo*cardium is *inside* the heart, the *myo*cardium is the *muscle* between the two, and the *peri*cardium surrounds the whole package. By the way, the root *cardi–* comes from the Greek word for heart, *kardia.*

The pericardium is made up of two layers — a tough, inelastic outer sac called the *fibrous pericardium* and an inner, serous *pericardium.* The serous pericardium is divided into two layers. The parietal pericardium is fused to the fibrous pericardium and the visceral pericardium is part of the epicardium (outer layer of the heart). The visceral layer extends to the vessels, where it merges with the parietal layer of the serous pericardium. Between the serous layers of the epicardium and the parietal pericardium are the small *pericardial space* and its tiny amount of lubricating *pericardial fluid.* This watery substance prevents irritation during *systole* (contraction of the heart) and *diastole* (relaxation of the heart).

Give these practice questions a try to see if you have the rhythm of all this:

1.–5. Match the description to its anatomical term.

1. _____ The system for gaseous exchange in the lungs

2. _____ The system for maintaining a constant internal environment in other tissues

3. _____ The membranous sac that surrounds the heart

4. _____ The wall that divides the heart into two cavities

5. _____ Uppermost two chambers of the heart

    a. Pericardium

    b. Pulmonary circuit

    c. Systemic circuit

    d. Atria

    e. Septum

6. What's the difference between the pulmonary and systemic circuits?

a. The pulmonary circuit supplies blood to the brain, and the systemic circuit supplies it everywhere else.

b. The pulmonary circuit supplies blood to the extremities, and the systemic circuit supplies blood to the internal organs.

c. The pulmonary circuit supplies blood to the skin, and the systemic circuit supplies blood to the underlying muscles.

d. The pulmonary circuit supplies blood to the left side of the body, and the systemic circuit supplies blood to the right side.

e. The pulmonary circuit travels to and from the lungs, and the systemic circuit keeps pressure and flow up for the body's tissues.

7. True or false: The heart is centrally located in the chest.

8. A closed system of circulation involves

    a. non-confinement of blood, general dispersal, minimal regulation.

    b. confinement of blood, general dispersal, critical regulation.

    c. non-confinement of blood, specific targeting, critical regulation.

    d. confinement of blood, specific targeting, critical regulation.

    e. confinement of blood, specific targeting, minimal regulation.

9. Only one of the heart wall's three layers contains muscle tissue. Which one?

    a. Pericardium

    b. Epicardium

    c. Myocardium

    d. Endocardium

    e. Endothelium

10.–14. Match the description to its anatomical term.

    10. _____ A membranous, serous layer attached to a fibrous sac

    11. _____ A tissue composed of layers and bundles of cardiac muscles

    12. _____ Outside layer of the heart wall that's interspersed with adipose

    13. _____ The interior lining of the heart

    14. _____ External grooves that indicate the regions of the heart

    a. Visceral pericardium

    b. Sulci

    c. Endocardium

    d. Myocardium

    e. Parietal pericardium

Finding the Key to the Heart's Chambers

The heart's lower two chambers, the ventricles, are quite a bit larger than the two atria up top. Yet the proper anatomical terms for their positions refer to the atria as being "superior" (above) and the ventricles "inferior" (below). In this case, size isn't the issue at all.

The atria

Sometimes referred to as "receiving chambers" because they receive blood returning to the heart through the veins, each atrium has two parts: a *principal cavity* with a smooth interior surface and an *auricle* (atrial appendage), a smaller, dog-ear-shaped pouch with muscular ridges inside called *pectinate muscles,* or *musculi pectinati,* that resemble the teeth of a comb. Some sources say that these musculi pectinati help strengthen the heart's contraction. They are less pronounced in the left auricle than in the right.

The right atrium appears slightly larger than the left and has somewhat thinner walls than the left. Its principal cavity, the *sinus venarum cavarum,* is between the two *vena cavae* and the *atrioventricular* (between an atrium and a ventricle) openings. The point where the right atrium's auricle joins with its principal cavity is marked externally by the *sulcus terminalis* and internally by the *crista terminalis*. Openings into the right atrium include the following:

✔ The *superior vena cava,* which has no valve and returns blood from the head, thorax, and upper extremities, directing it toward the atrioventricular opening.

✔ The *inferior vena cava,* which returns blood from the trunk and lower extremities and directs it toward the *fossa ovalis* in the interatrial septum, which also has no valve.

✔ The *coronary sinus,* which opens between the inferior vena cava and the atrioventricular opening, returns blood from the heart, and is covered by the ineffective *Thebesian valve.*

✔ An atrioventricular opening covered by the *tricuspid valve (AV valve).*

✔ The *fossa ovalis,* an oval depression in the interatrial septum that corresponds to the foramen ovale of the fetal heart.

The left atrium's principal cavity contains openings for the four *pulmonary veins,* two from each lung, which have no valves. Frequently, the two left veins share a common opening. The left auricle, a dog-ear-shaped blind pouch, is longer, narrower, and more curved than the right, marked interiorly by the pectinate muscles, and the left atrium's atrioventricular (AV) opening is smaller than on the right and is protected by the *mitral,* or *bicuspid,* valve (AV valve).

The ventricles

The heart's ventricles are sometimes called the pumping chambers because it's their job to receive blood from the atria and pump it to the lungs and out into the body's network of arteries. More force is needed to move the blood great distances, so the myocardium of the ventricles is thicker than that of either atrium, and the myocardium of the left ventricle is thicker than that of the right.

The right ventricle only has to move blood to the lungs, so its myocardium is only one-third as thick as that of its neighbor to the left. Roughly triangular in shape, the right ventricle occupies much of the *sternocostal* (front) surface of the heart and forms the *conus arteriosus* where it joins the *pulmonary trunk.* The right ventricle extends downward toward where the heart rests against the diaphragm. A circular opening into the pulmonary trunk is covered by the *pulmonary semilunar valve,* so-called because of its three crescent-shaped cusps. When the ventricle relaxes, the blood from the pulmonary artery tends to flow back toward the ventricle, filling the pockets of the cusps and causing the valve to close. The oval AV opening is surrounded by a strong fibrous ring and covered by the tricuspid valve (AV valve), named after its three unequally sized cusps. The atrial surface of the tricuspid valve is smooth, but the side toward the ventricle is irregular, forming a ragged edge where the *chordae tendineae* attach. These fibrous cords, which are attached to nipple-shaped projections called *papillary muscles* in the ventricle's wall, prevent blood from flowing back into the atrium. Cardiac muscle in the ventricle's wall is in an irregular pattern of bundles and bands called the *trabeculae carneae.*

Longer and more conical in shape, the left ventricle's tip forms the apex of the heart. Its walls are three times thicker than those of the right ventricle. Its AV opening is smaller than that in the right ventricle and is covered by the bicuspid, or mitral, valve (AV valve) that's comprised of two unequal cusps. This ventricle's chordae tendineae are fewer, thicker, and stronger, and they're attached by only two larger papillary muscles, one on the front (anterior) wall and one on the back (posterior). More ridges are packed more densely in the muscular trabeculae carneae. Its opening to the aorta is protected by the *aortic semilunar valve,* composed of three half-moon cusps that are larger, thicker, and stronger than the pulmonary (semilunar) valve's cusps. The blood enters the ascending aorta into the aortic arch where three vessels branch off:

- The brachiocephalic trunk that will divide into the right common carotid artery and the right subclavian artery
- The left common carotid artery
- The left subclavian artery

Between these cusps and the aortic wall are dilated pockets called *aortic sinuses,* which are openings for the *coronary arteries,* which carry blood to the cardiac tissue. The blood is returned from the apex of the heart through the great (left) cardiac veins, which enter the right atrium through the coronary sinus.

Pump up your practice time with these questions related to the chambers of the heart:

15.–30. Use the terms that follow to identify the heart's major vessels shown in Figure 10-2.

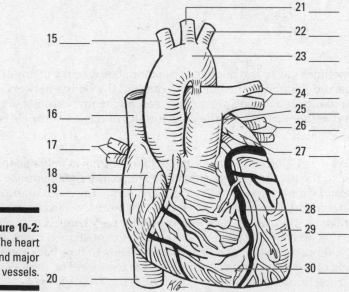

Figure 10-2:
The heart
and major
vessels.

Illustration by Kathryn Born, MA

a. Left pulmonary veins

b. Left ventricle

c. Brachiocephalic trunk

d. Right pulmonary veins

e. Right ventricle

f. Left subclavian artery

g. Right coronary artery

    h. Aortic arch

    i. Great (left) cardiac vein

    j. Superior vena cava

    k. Left common carotid artery

    l. Left pulmonary arteries

    m. Left atrium

    n. Inferior vena cava

    o. Pulmonary trunk

    p. Right atrium

31.–34. Use the terms that follow to identify the heart valves shown in Figure 10-3. (**Note:** In a beating heart, either the two top or the two bottom valves would be open whenever the opposite pair is closed. Figure 10-3, however, gives a better view of all four valves simultaneously.)

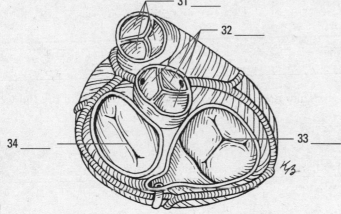

31 _____

32 _____

33 _____

34 _____

Figure 10-3:
The heart
valves.

Illustration by Kathryn Born, MA

    a. Tricuspid valve

    b. Pulmonary semilunar valve

    c. Aortic semilunar valve

    d. Biscuspid valve

35. Why is the myocardium of the ventricles thicker than the myocardium of the atria?

    a. The myocardium keeps valves in place, and the ventricular valves have to hold back greater blood pressure with each contraction.

    b. The myocardium is the inner lining of the heart, and the ventricles must hold greater volumes of blood between contractions.

    c. The myocardium is the muscular tissue of the heart, and more force is needed in chambers pumping blood greater distances.

    d. The myocardium consists mostly of connective tissue, and the ventricles are more intricate than the atria.

36. The cavity in the heart that contains the areas called the sinus venarum cavarum and a blind pouch called the auricle is the

a. left ventricle.

b. right atrium.

c. left atrium.

d. right ventricle.

37. The superior vena cava enters the heart by way of the

a. left ventricle.

b. right ventricle.

c. left atrium.

d. right atrium.

38. What holds the cusps of the atrioventricular valves in place?

a. Supporting ligaments

b. The chordae tendineae

c. The trabeculae carneae

d. The papillary muscles

39. What covers the atrioventricular opening between the right atrium and the right ventricle?

a. The bicuspid valve

b. The tricuspid valve

c. Pulmonary semilunar valve

d. The aortic semilunar valve

40.–44. Match the following descriptions with the proper anatomical terms.

40. _____ Returns blood to the heart from the head, thorax, and upper extremities

41. _____ Valve located between the right atrium and right ventricle

42. _____ Valve located between the right ventricle and pulmonary trunk

43. _____ Returns blood to the heart from the trunk and lower extremities

44. _____ Valve located between the left atrium and left ventricle

a. Tricuspid valve (AV valve)

b. Bicuspid valve (AV valve)

c. Superior vena cava

d. Pulmonary semilunar valve

e. Inferior vena cava

Conducting the Heart's Music

The mighty, nonstop heart keeps up its rhythm because of a carefully choreographed dance of electrical impulses called the *conduction system* that has the power to produce a spontaneous rhythm and conduct an electrical impulse. Five structures play key roles in this dance — the *sinoatrial node, atrioventricular node, atrioventricular bundle (bundle of His), bundle branches,* and *Purkinje fibers.* (You can see them in Figure 10-4.) Each is formed of highly tuned, modified cardiac muscle. Rather than both contracting and conducting impulses as other cardiac muscle does, these structures specialize in conduction alone, setting the pace for the rest of the heart. Following is a bit more information about each one:

✔ **Sinoatrial node:** This node really is the pacemaker of the heart. Located at the junction of the superior vena cava and the right atrium, this small knot, or mass, of specialized cardiac muscle cells initiates an electrical impulse that moves over the musculature of both atria, causing atrial walls to contract simultaneously and emptying blood into both ventricles. It's also called the *S-A node, sinoauricular node,* and *sinus node.*

✔ **Atrioventricular node:** The impulse that starts in the S-A node moves to this mass of modified cardiac tissue that's located in the septal wall of the right atrium. Also called the *A-V node,* it directs the impulse to the A-V bundles in the septum.

✔ **Atrioventricular bundle:** From the A-V node, the impulse moves into the atrioventricular bundle, also known as the *A-V bundle* or *bundle of His* (pronounced "hiss").

✔ **Bundle branches:** The atrioventricular bundle breaks into two branches that extend down the sides of the interventricular septum under the endocardium to the heart's apex.

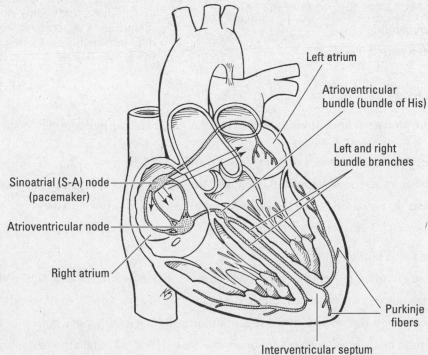

Figure 10-4: The conductive system of the heart.

Illustration by Kathryn Born, MA

✔ **Purkinje fibers:** At the apex, the bundles break up into *terminal conducting fibers,* or Purkinje fibers, and merge with the muscular inner walls of the ventricles. The pulse then stimulates ventricular contraction that begins at the apex and moves toward the base of the heart, forcing blood toward the aorta and pulmonary trunk.

One of the best ways to detect cardiac tissue under a microscope is to look for undulating double membranes called *intercalated discs* separating adjacent cardiac muscle fibers. *Gap junctions* in the discs permit ions to pass between the cells, spreading the *action potential* of the electrical impulse and synchronizing cardiac muscle contractions. Potential problems include *fibrillation,* a breakdown in rhythm or propagation of the impulses that causes individual fibers to act independently, and *heart block,* an interruption that causes the atria and ventricles to take on their own rates of contraction. Usually the atria contract faster than the ventricles.

A healthy heart makes a "lub-dub" sound as it beats. The first sound (the "lub") is heard most clearly near the apex of the heart and comes at the beginning of *ventricular systole* (the closing of the atrioventricular valves and opening of the semilunar valves). It's lower in pitch and longer in duration than the second sound (the "dub"), heard most clearly over the second rib, which results from the semilunar valves closing during ventricular diastole. Defects in the valves can cause turbulence or regurgitation of blood that can be heard through a stethoscope. Called *murmurs,* these sounds indicate imperfect closure of one or more valves.

Have you got the beat? Try the following practice questions that deal with the heart's rhythm:

Q. Cardiac tissue is distinctive microscopically because of the presence of

  a. hemoglobin.

  b. intercalated discs.

  c. fibrin.

  d. ganglia.

A. The correct answer is intercalated discs. These undulating double membranes separate adjacent cardiac muscle fibers, so they're unique to cardiac tissue.

45. Why is the sinoatrial node (also known as the S-A node, sinoauricular node, or sinus node) called the heart's "pacemaker"?

  a. It interrupts irregular heart rhythms to set a steadier pace.

  b. It initiates an electrical impulse that causes atrial walls to contract simultaneously.

  c. It speeds up and slows down the heart, depending on the body's activities.

  d. It supplies a boost of energy to jump-start the heart when it slows.

46. What is a heart murmur?

  a. A spasm of cardiac muscle that causes an extra sound between the heart's "lub" and its "dub"

  b. An extra pulse in a heart contraction that mutes the heart's "lub" sound

  c. The sound of turbulence or blood regurgitation resulting from a valve defect

  d. The sound made when a section of heart muscle remains still while the rest of the heart contracts around it

47. Choose the correct conductive pattern.

   a. S-A node → Bundle of His → bundle branches → A-V node → Purkinje fibers

   b. A-V node → S-A node → Purkinje fibers → Bundle of His → bundle branches

   c. S-A node → A-V node → Bundle of His → bundle branches → Purkinje fibers

   d. S-A node → Purkinje fibers → Bundle of His → bundle branches → A-V node

Riding the Network of Blood Vessels

Blood vessels come in three varieties, which you can see illustrated in Figure 10-5:

- **Arteries:** *Arteries* carry blood away from the heart; all arteries, except the pulmonary arteries, carry oxygenated blood. The largest artery is the aorta. Small ones are called *arterioles,* and microscopically small ones are called *metarterioles.*

- **Veins:** *Veins* carry blood toward the heart; all veins except the pulmonary veins contain deoxygenated blood. Small ones are called *venules,* and large venous spaces are called *sinuses.*

- **Capillaries:** Microscopically small *capillaries* carry blood from arterioles to venules, but sometimes tiny spaces in the liver and elsewhere, called *sinusoids,* replace capillaries.

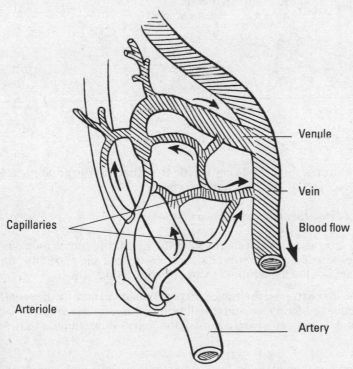

Figure 10-5: The capillary exchange.

Venule

Vein

Blood flow

Capillaries

Arteriole

Artery

Illustration by Kathryn Born, MA

The walls of arteries and veins have three layers (see Figure 10-6):

✔ The outermost *tunica externa* (sometimes called *tunica adventitia*) composed of white fibrous connective tissue

✔ A central "active" layer called the *tunica media* composed of smooth muscle fibers and yellow elastic fibers

✔ An inner layer called the *tunica intima* made up of endothelium that aids in preventing blood coagulation by reducing the resistance of blood flow

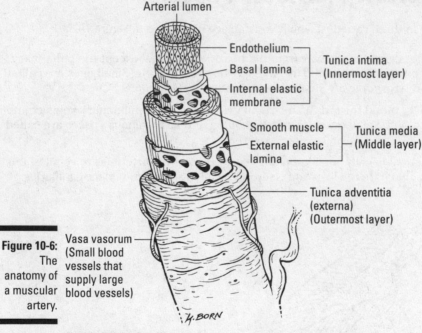

Arterial lumen

Endothelium

Basal lamina

Internal elastic membrane

Tunica intima (Innermost layer)

Smooth muscle

External elastic lamina

Tunica media (Middle layer)

Tunica adventitia (externa) (Outermost layer)

Figure 10-6: The anatomy of a muscular artery.

Vasa vasorum (Small blood vessels that supply large blood vessels)

Illustration by Kathryn Born, MA

Arterial walls are very strong, thick, and very elastic to withstand the great pressure to which the arteries are subjected. Arteries have no valves.

There are two types of arteries: elastic and muscular:

✔ **Elastic:** In elastic arteries, found primarily near the heart, the tunica media is composed of yellow elastic fibers that stretch with each systole and recoil during diastole; essentially they act as shock absorbers to smooth out blood flow.

✔ **Muscular:** In muscular arteries, the tunica media consists primarily of smooth muscle fibers that are active in blood flow and distribution of blood. The larger blood vessels have smaller blood vessels, the *vasa vasorum,* that carry nourishment to the vessel wall itself.

Although larger in diameter than arteries, veins have thinner walls and are less distensible and elastic. Veins that carry blood against the force of gravity, such as those in the legs and feet, contain valves to prevent backflow into the capillaries. Normally the blood that veins are returning to the heart is deoxygenated (contains carbon dioxide); the one exception is the pulmonary vein, which returns oxygenated blood to the heart from the lungs.

All gaseous exchange occurs by diffusion across the capillary walls. Capillaries are breathtakingly tiny and capable of forming vast networks, or capillary beds. Precapillary sphincters take the place of valves to regulate blood flow. Capillary walls are a single layer of squamous endothelial cells to facilitate the oxygen and carbon dioxide exchange with tissues.

The following sections cover two distinct types of circulation: hepatic and fetal.

Hepatic circulation

Blood from the digestive tract takes a detour through the *hepatic portal vein* to the liver before continuing on to the heart. Called the *hepatic portal system,* this circuitous route helps regulate the amount of glucose circulating in the bloodstream (see Figure 10-7). As the blood flows through the sinusoids of the liver, *hepatic parenchymal cells* remove the nutrient materials. *Phagocytic (Kupffer)* cells in the sinusoids remove bacteria and other foreign materials from the blood. The blood exits the liver by the hepatic veins, which carry it to the inferior vena cava, which ultimately returns it to the heart.

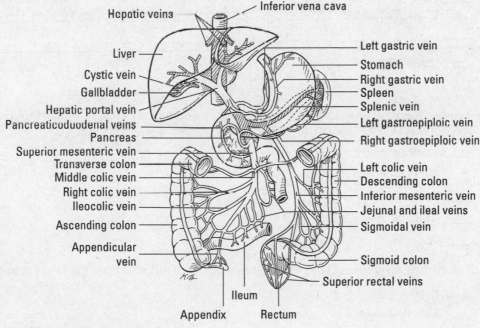

Figure 10-7:
The veins of the hepatic portal system.

Illustration by Kathryn Born, MA

Fetal circulation

Because nutrients and oxygen come from the mother's bloodstream, fetal circulation requires extra vessels to get the job done. Two umbilical arteries — the *umbilical vein* and the *ductus venosus* — fill the bill. Fetal blood leaves the placenta through the umbilical vein, which branches at the liver to become the *ductus venosus* before entering the inferior vena cava that carries blood to the right atrium and then through a hole in the septum called the *foramen ovale* into the left atrium. From there it flows into the left ventricle and is pumped through the aorta to the head, neck, and upper extremities. It returns to the heart through the superior vena cava, to the right atrium, to the right ventricle, to the pulmonary trunk (lungs inactive), goes through the *ductus arteriosus* into the aorta, to the abdominal and pelvic viscera and lower extremities, and to the placenta through the umbilical artery. After birth, these circulation pathways quickly shut down, eventually leaving a depression in the septum, the fossa ovalis, where the foramen ovale once was.

Now is your chance to practice circulating through the circulatory system:

48. Why does a fetus have extra blood vessels?

a. As a countermeasure against possible infections coming from the mother

b. To move nutrients and oxygen from the mother's bloodstream into fetal circulation

c. To circumnavigate the foramen ovale

d. To provide additional circulation to the ductus venosus

49. Follow a drop of blood through the heart, starting in the superior vena cava. Number the following structures in sequential order.

_____ Pulmonary vein

_____ Right ventricle

_____ Lung

_____ Right atrium

_____ Pulmonary artery

50. Number the structures in the correct sequence of blood flow from the heart to the radial artery for pulse. Start at the heart with the aortic semilunar valve.

_____ Axillary artery

_____ Subclavian artery

_____ Ascending aorta

_____ Brachial artery

_____ Aortic arch

51. Number the structures in the correct sequence of blood flow from the forearm to the heart.

_____ Basilic vein

_____ Subclavian vein

_____ Superior vena cava

_____ Brachial vein

_____ Axillary vein

52. Number the structures in the correct sequence of blood flow from the great saphenous vein back to the heart.

_____ External iliac vein

_____ Right atrium

_____ Common iliac vein

_____ Femoral vein

_____ Inferior vena cava

53. Follow a drop of blood from the right atrium to the radial artery (for pulse). Number the structures in sequential order.

__1__ Right atrium

_____ Pulmonary artery

_____ Left atrium

_____ Bicuspid valve (AV valve)

_____ Ascending aorta

_____ Axillary artery

_____ Radial artery

_____ Right ventricle

_____ Pulmonary vein

_____ Aortic semilunar valve

_____ Brachial artery

_____ Tricuspid valve (AV valve)

_____ Lung capillary

_____ Aortic arch

_____ Pulmonary semilunar valve

_____ Left ventricle

_____ Subclavian artery

54. Follow a drop of blood from the stomach to the inferior vena cava. (Remember the portal system?) Number the structures in sequential order.

__1__ Superior mesenteric vein

_____ Sinusoids of liver

_____ Hepatic vein

_____ Hepatic portal vein

_____ Inferior vena cava

55. Follow a drop of blood from the aortic semilunar valve of the heart to the forearm and back to the heart. Number the structures in sequential order.

___1___ Ascending aorta

_____ Basilic vein

_____ Axillary artery

_____ Subclavian vein

_____ Brachial vein

_____ Radial artery

_____ Right atrium

_____ Brachial artery

_____ Axillary vein

_____ Capillaries in the hand

_____ Superior vena cava

_____ Subclavian artery

_____ Aortic arch

56. Follow a drop of blood from the saphenous vein back to the heart. Number the structures in sequential order.

___1___ Saphenous vein

_____ External iliac vein

_____ Inferior vena cava

_____ Right atrium

_____ Femoral vein

_____ Common iliac vein

_____ Right ventricle

57. Follow a drop of blood from the anterior tibial vein to the lungs. Number the structures in sequential order.

___1___ Anterior tibial vein

_____ External iliac vein

_____ Inferior vena cava

_____ Right ventricle

_____ Popliteal vein

_____ Common iliac vein

_____ Pulmonary artery

_____ Femoral vein

_____ Right atrium

_____ Lung capillaries

Answers to Questions on the Circulatory System

The following are answers to the practice questions presented in this chapter.

1 The system for gaseous exchange in the lungs: **b. Pulmonary circuit**

2 The system for maintaining a constant internal environment in other tissues: **c. Systemic circuit**

3 The membranous sac that surrounds the heart: **a. Pericardium**

4 The wall that divides the heart into two cavities: **e. Septum**

5 Uppermost two chambers of the heart: **d. Atria**

6 What's the difference between the pulmonary and systemic circuits? **e. The pulmonary circuit travels to and from the lungs, and the systemic circuit keeps pressure and flow up for the body's tissues.** The Latin *pulmon–* means "lung." None of the other answers even mentions the lungs.

7 True or false: The heart is centrally located in the chest. **False.** Actually, two-thirds of the heart lies to the left of the body's center.

8 A closed system of circulation involves **d. confinement of blood, specific targeting, critical regulation.** In short, the system confines, targets, and regulates.

9 Only one of the heart wall's three layers contains muscle tissue. Which one? **c. Myocardium.** *Myo–* means "muscle," *epi–* is "upon," *endo–* is "within," and *peri–* is "around."

10 A membranous, serous layer attached to a fibrous sac: **c. Parietal pericardium**

11 A tissue composed of layers and bundles of cardiac muscles: **d. Myocardium**

12 Outside layer of the heart wall that's interspersed with adipose: **a. Visceral pericardium**

13 The interior lining of the heart: **c. Endocardium**

14 External grooves that indicate the regions of the heart: **b. Sulci**

15–30 Following is how Figure 10-2, the heart, should be labeled.

15. **c. Brachiocephalic trunk;** 16. **j. Superior vena cava;** 17. **d. Right pulmonary veins;** 18. **p. Right atrium;** 19. **g. Right coronary artery;** 20. **n. Inferior vena cava;** 21. **k. Left common carotid artery;** 22. **f. Left subclavian artery;** 23. **h. Aortic arch;** 24. **l. Left pulmonary arteries;** 25. **o. Pulmonary trunk;** 26. **a. Left pulmonary veins;** 27. **m. Left atrium;** 28. **i. Great (left) cardiac vein;** 29. **b. Left ventricle;** 30. **e. Right ventricle**

31–34 Following is how Figure 10-3, the heart valves, should be labeled.

31. **b. Pulmonary semilunar valve;** 32. **c. Aortic semilunar valve;** 33. **a. Tricuspid valve;** 34. **d. Bicuspid valve**

35 Why is the myocardium of the ventricles thicker than the myocardium of the atria? **c. The myocardium is the muscular tissue of the heart, and more force is needed in chambers pumping blood greater distances.** The correct answer is the only one that correctly identifies myocardium as muscle tissue. Remember: *myo–* means "muscle."

36 The cavity in the heart that contains the areas called the sinus venarum cavarum and a blind pouch called the auricle is the **b. right atrium.**

37 The superior vena cava enters the heart by way of the **d. right atrium.**

38 What holds the cusps of the atrioventricular valves in place? **b. The chordae tendineae.**

39 What covers the atrioventricular opening between the right atrium and the right ventricle? **b. The tricuspid valve.** Sorry if this seemed to be a trick question, but even if you have trouble remembering the heart's right openings from its left ones, you simply need to remember that the bicuspid and the mitral valve are the same thing, so "tricuspid valve" is the only correct answer here.

40 Returns blood to the heart from the head, thorax, and upper extremities: **c. Superior vena cava**

41 Valve located between the right atrium and right ventricle: **a. Tricuspid valve (AV valve)**

42 Valve located between the right ventricle and pulmonary trunk: **d. Pulmonary semilunar valve**

43 Returns blood to the heart from the trunk and lower extremities: **e. Inferior vena cava**

44 Valve located between the left atrium and left ventricle: **b. Bicuspid valve (AV valve)**

45 Why is the sinoatrial node (also known as the S-A node, sinoauricular node, or sinus node) called the heart's "pacemaker"? **b. It initiates an electrical impulse that causes atrial walls to contract simultaneously.** The other answers address rhythm, but the correct answer shows how the S-A node coordinates the heart's contractions.

46 What is a heart murmur? **c. The sound of turbulence or blood regurgitation resulting from a valve defect.** The correct answer is the only one that relates a murmur to a valve defect.

47 Choose the correct conductive pattern. **c. S-A node → A-V node → Bundle of His → bundle branches → Purkinje fibers.**

48 Why does a fetus have extra blood vessels? **b. To move nutrients and oxygen from the mother's bloodstream into fetal circulation.**

49 Follow a drop of blood through the heart, starting in the superior vena cava. Number the structures in sequential order. **1. Right atrium; 2. Right ventricle; 3. Pulmonary artery; 4. Lung; 5. Pulmonary vein.**

50 Number the structures in the correct sequence of blood flow from the heart to the radial artery for pulse. Start at the heart with the aortic semilunar valve. **1. Ascending aorta; 2. Aortic arch; 3. Subclavian artery; 4. Axillary artery; 5. Brachial artery.**

51 Number the structures in the correct sequence of blood flow from the forearm to the heart. **1. Basilic vein; 2. Brachial vein; 3. Axillary vein; 4. Subclavian vein; 5. Superior vena cava.**

52 Number the structures in the correct sequence of blood flow from the great saphenous vein back to the heart. **1. Femoral vein; 2. External iliac vein; 3. Common iliac vein; 4. Inferior vena cava; 5. Right atrium.**

53 Follow a drop of blood from the right atrium to the radial artery (for pulse). Number the structures in sequential order. **1. Right atrium; 2. Tricuspid valve (AV valve); 3. Right ventricle; 4. Pulmonary semilunar valve; 5. Pulmonary artery; 6. Lung capillary; 7. Pulmonary vein; 8. Left atrium; 9. Bicuspid valve (AV valve); 10. Left ventricle; 11. Aortic semilunar valve; 12. Ascending aorta; 13. Aortic arch; 14. Subclavian artery; 15. Axillary artery; 16. Brachial artery; 17. Radial artery.**

54 Follow a drop of blood from the stomach to the inferior vena cava. (Remember the portal system?) Number the structures in sequential order. **1. Superior mesenteric vein; 2. Hepatic portal vein; 3. Sinusoids of liver; 4. Hepatic vein; 5. Inferior vena cava.**

55 Follow a drop of blood from the aortic semilunar valve of the heart to the forearm and back to the heart. Number the structures in sequential order. **1. Ascending aorta; 2. Aortic arch; 3. Subclavian artery; 4. Axillary artery; 5. Brachial artery; 6. Radial artery; 7. Capillaries in the hand; 8. Basilic vein; 9. Brachial vein; 10. Axillary vein; 11. Subclavian vein; 12. Superior vena cava; 13. Right atrium.**

56 Follow a drop of blood from the saphenous vein back to the heart. Number the structures in sequential order. **1. Saphenous vein; 2. Femoral vein; 3. External iliac vein; 4. Common iliac vein; 5. Inferior vena cava; 6. Right atrium; 7. Right ventricle.**

57 Follow a drop of blood from the anterior tibial vein to the lungs. Number the structures in sequential order. **1. Anterior tibial vein; 2. Popliteal vein; 3. Femoral vein; 4. External iliac vein; 5. Common iliac vein; 6. Inferior vena cava; 7. Right atrium; 8. Right ventricle; 9. Pulmonary artery; 10. Lung capillaries.**

Chapter 11

Keeping Up Your Defenses: The Lymphatic System

You see it every rainy day — water, water everywhere, rushing along gutters and down storm drains into a complex underground system that most would rather not give a second thought. Well, it's time to give hidden drainage systems a second thought: Your body has one, called *the lymphatic system*. This chapter explains what you need to know about this system.

Duct, Duct, Lymph

You already know that the body is made up mostly of fluid. *Interstitial* or *extracellular fluid* moves in and around the body's tissues and cells constantly. It leaks out of blood capillaries at the rate of nearly 51 pints a day, carrying various substances to and away from the smallest nooks and crannies. Most of that fluid gets reabsorbed into blood capillaries. But the one or two liters of extra fluid that remain around the tissues become a substance called *lymph* that needs to be managed to maintain fluid balance in the internal environment. That's where the lymphatic system steps in, forming an alternative route for the return of tissue fluid to the bloodstream.

But the lymphatic system is more than a drainage network. It's a body-wide filter that traps and destroys invading microorganisms as part of the body's immune response network. It can remove impurities from the body, help absorb and digest excess fats, and maintain a stable blood volume despite varying environmental stresses. Without it, the cardiovascular system would grind to a halt.

The story of the lymphatic system (shown in Figure 11-1) begins deep within the body's tissues at the farthest reaches of blood capillaries, where nutrients, plasma, and plasma proteins move out into tissues, while waste products like carbon dioxide and the fluid carrying those molecules move back in through a process known as *diffusion*. Roughly 10 percent of the fluid that leaves the capillaries remains deep within the tissues as part of the interstitial (meaning "between the tissues") fluid. But in order for the body to maintain a sufficient volume of water within the circulatory system, eventually this interstitial fluid and its protein must get back into the blood. So the lymphatic vessels act as a recycling system to gather, transport, cleanse, and return this fluid to the bloodstream.

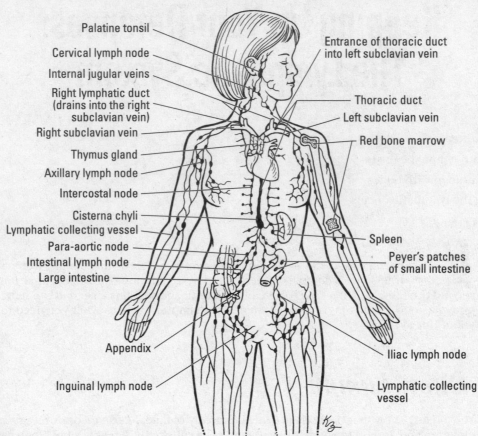

Figure 11-1:
The lymphatic system.

Illustration by Kathryn Born, MA

To collect the fluid, minute vessels called *lymph capillaries* are woven throughout the body, with a few caveats and exceptions. There are no lymph capillaries in the central nervous system, teeth, outermost layer of the skin, certain types of cartilage, any other avascular tissue, and bones. And because bone marrow makes lymphocytes, which we explain in the next section, it's considered part of the lymphatic system. Plus, *lacteals* (lymphatic capillaries found in the villi of the intestines) absorb fats to mix with lymph, forming a milky fluid called *chyle.* (See Chapter 9 for details on lacteals.)

Unlike blood capillaries, lymph capillaries dead-end (terminate) within tissue. Made up of a single layer of loosely overlapping endothelial cells anchored by fine filaments, lymph capillaries behave as if their walls are made of cellular, one-way valves. Endothelium provides a smooth, slick, friction-free surface in the capillaries. When the pressure outside a capillary is greater than it is inside, the filaments anchoring the cells allow them to open, permitting interstitial fluid to seep in. Increasing pressure inside the capillary walls eventually forces the cell junctions to close. Once in the capillaries, the trapped fluid is known as *lymph,* and it moves into larger, veinlike *lymphatic vessels.* The lymph moves slowly and without any kind of central pump through a combination of peristalsis, the action of semilunar valves, and the squeezing influence of surrounding skeletal muscles, much like that which occurs in veins.

In the deeper layers of the skin, lymph vessels form networks around veins, but in the trunk of the body and around internal organs, they form networks around arteries. Lymph vessels have thinner walls than veins, are wider, have more valves, and — most important — regularly bulge with bean-shaped sacs called *lymph nodes* (more on those in the next section). Just as small tree branches merge into larger ones and then into the trunk, *lymphatics* eventually merge into the nine largest lymphatic vessels called *lymphatic trunks.* The biggest of these at nearly 1½ feet in length is the *thoracic duct;* nearly all the body's lymph vessels empty into it. Only those vessels in the right half of the head, neck, upper extremity, and thorax empty into its smaller mate, the *right lymphatic duct.* Lymph returns to the bloodstream when both ducts connect with the *subclavian veins* (under the collarbone).

The thoracic duct, which also sometimes is called the *left lymphatic duct,* arises from a triangular sac called the *chyle cistern* (or *cisterna chyli*) into which one *intestinal trunk* and two *lumbar lymphatic trunks* (which drain the lower limbs) flow. Both the thoracic duct and the much smaller right lymphatic duct drain into the subclavian veins. The remaining four trunks are a pair serving the *jugular* region (sides of the throat) and a pair serving the *bronchomediastinal* region (the central part of the chest).

To see how much of this information is seeping in, answer the following questions:

1. The lymphatic system plays an important role in regulating

   a. intracellular energy function.

   b. interstitial fluid protein.

   c. metabolizing unused fats from the small intestine.

   d. intercellular transportation of oxygen.

2. Under what circumstances do the walls of lymphatic capillaries open?

   a. When the filaments holding the cells together begin to dehydrate

   b. When the pressure outside the capillaries is greater than inside

   c. When the capillaries sense extra salts in the bloodstream

   d. When reticular fibers begin to force their way between the cells

3. How does lymph move around inside its vessels?

   a. Rapidly, through the action of tiny pumps called lymph nodes

   b. Slowly, and only as additional lymph enters the system

   c. Slowly, through peristalsis, the action of semilunar valves, and the squeezing of surrounding skeletal muscles

   d. Steadily and rhythmically, mirroring the heart rate

4. What is the largest lymphatic vessel in the body?

   a. Right lymphatic duct

   b. Spleen

   c. Thoracic duct

   d. Chyle cistern

5. The lymphatic system serves several crucial functions, but what does it *not* normally do?

a. Return interstitial fluid to the blood

b. Destroy bacteria

c. Remove old erythrocytes

d. Produce erythrocytes

Poking at the Nodes

Lymph nodes (see Figure 11-2) are the site of filtration of the lymphatic system. Also sometimes incorrectly referred to as *lymph glands* — they don't secrete anything, so technically they're not glands — these kidney-shaped sacs are surrounded by connective tissue (and therefore are tough to spot). Lymph nodes contain macrophages, which destroy bacteria, cancer cells, and other matter in the lymph fluid. *Lymphocytes,* which produce an immune response to microorganisms, also are found in lymph nodes. Like the kidneys, the indented part of each node is referred to as the *hilus* (see Chapter 12 for details). The *stroma* (body) of each node is surrounded by a fibrous capsule that dips into the node to form *trabeculae,* or *septa* (thin dividing walls) that divide the node into compartments. *Reticular* (netlike) fibers are attached to the trabeculae and form a framework for the lymphoid tissue and *lymphocytes* (white blood cells) in clusters called *lymphatic nodules.*

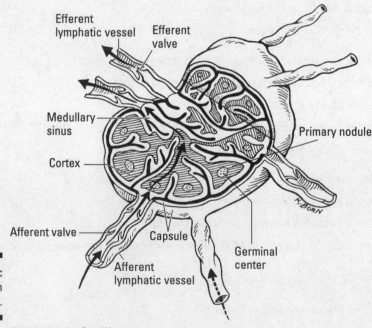

Figure 11-2:
A lymph
node.

Illustration by Kathryn Born, MA

Inside the node is a *cortex* where most of the lymphocytes gather, and at the center is a *medulla,* which is less dense than the cortex but also contains lymphocytes. The medulla is arranged into elongated masses known as *medullary cords*. Medullary cords are thin inward extensions of cortical lymphoid tissue (T and B lymphocytes) bounded by lymph sinuses in the medulla. The medullary cords also contain antibody-secreting plasma cells, macrophages, B and T lymphocytes, plasmacytoid lymphocytes, and plasmablasts. The outer cortex consists of lymphocytes arranged in masses called *lymphatic nodules,* which have central areas called *germinal centers* that produce the lymphocytes. Lymph fluid enters the node on its convex side through *afferent* (inbound) lymphatic vessels that have valves opening only toward the node. Lymph circulates through the node, where it's filtered and then allowed to depart through *efferent* (outbound) lymphatic vessels in the hilus with valves pointing exclusively away from the node.

If you have trouble remembering your afferent from your efferent, think of the "a" as standing for "access" and the "e" as standing for "exit."

Although some lymph nodes are isolated from others, most nodes occur in groups, or clusters, particularly in the inguinal (groin), axillary (armpit), and mammary gland areas. (You can see some lymph nodes in Figure 11-1.) The following are the primary lymph node regions:

- **Head:** Nodes in the head drain lymph from the scalp, upper neck, ear, parts of the eye, nose, and cheek.

- **Neck:** Lymph vessels from the head and the neck carry lymph to the nodes in the neck.

- **Upper extremities:** Axillary nodes found in the armpit receive lymph from the upper arm, while the lymphatic vessels on the radial side of the arm supply nodes in the clavicle region.

- **Lower extremities:** The inguinal nodes function to drain the lower extremities.

- **Abdomen and pelvis:** Lymph nodes in the abdominal and pelvic region filter fluid from the lower body regions, reproductive organs, and thighs.

- **Viscera:** Viscera nodes (or gastric lymph nodes) function in the drainage of the digestive organs.

- **Thorax:** Lymph nodes in the chest region process lymph from the thoracic wall.

Note: There are a few exceptions to areas covered by the lymphatic vessels. The entire central nervous system, bone, teeth, and bone marrow contain no lymph vessels or nodes.

Each node acts like a filter bag filled with a network of thin, perforated sheets of tissue — a bit like cheesecloth — through which lymph must pass before moving on. White blood cells line the sheets of tissue, including several types that play critical roles in the body's immune defenses. This filtering action explains why, when infection first starts, lymph nodes often swell with the cellular activity of the immune system launching into battle with the invading microorganisms.

The cortex of each lymph node contains monocytes and two types of lymphocytes: *B cells* and *T cells.*

- **Monocytes** within the lymph nodes develop into large invader-eaters called *macrophages* that are capable of destroying a variety of microorganisms and sometimes even cancer cells.

✔ **B cells** don't attack pathogens directly but instead may produce molecules called *anti-bodies* that do the dirty work. Or they may instruct other cells called *phagocytes* (literally "cells that eat") to attack the invaders.

✔ **T cells** are lymphocytes that started out in the bone marrow but matured in the *thymus gland* (hence the name T cells) before moving on to the lymph nodes and spleen. (See the later section "T cell central: The thymus gland" for more information on T cells.)

Think you have a node-tion (sorry!) about what's happening here? Test your knowledge:

Q. An area of the body where no lymph nodes are found is the

   a. integument.

   b. liver.

   c. stomach.

   d. central nervous system (brain and spinal cord).

A. The correct answer is the central nervous system (brain and spinal cord). The central nervous system contains no lymph vessels or nodes.

6.–7. Fill in the blanks to complete the following sentence.

Lymph moves from a/an **6.**_____ vessel through a lymph node and back out into the blood through a/an **7.**_____ vessel.

8. What do lymph nodes use to destroy bacteria, cancer cells, and other foreign material?

   a. Stroma

   b. Trabeculae

   c. Peyer's patches

   d. Macrophages

9. Why do lymph nodes sometimes swell when an infection is present?

   a. To make room for additional lymph fluid

   b. Because of increased viscera that results from gastric activity

   c. To accommodate cellular activity of the immune-system-attacking microorganisms

   d. Because white blood cells are larger than red blood cells

10. Lymph nodes have two primary functions. What are they?

   a. To conserve white blood cells and to produce lymphocytes

   b. To produce bilirubin and to cleanse lymph fluid

   c. To produce lymphocytes and antibodies and to filter lymph fluid

   d. To conserve iron and to remove erythrocytes

11. What type of cell can be found in a lymph node's medulla?

   a. T lymphocytes, also known as T cells

   b. Monocytes that can develop into macrophages

   c. B lymphocytes, also known as B cells

   d. Both a and c

12. The germinal center of lymph nodules produces what type of cells?

   a. Monocytes

   b. Lymphocytes

   c. Phagocytes

   d. Trabeculae

13. The connective tissue fiber that forms the framework of the lymphoid tissue is

   a. cartilaginous.

   b. collagenous.

   c. reticular.

   d. elastic.

14. What shape is a lymph node?

   a. Rectangular like a shoe box

   b. Flat like a disc

   c. Round like a ping-pong ball

   d. Oval and slightly bent, like a kidney bean

15. Why are T cells called by that name?

   a. They are shaped like the letter T.

   b. They are produced in the thyroid.

   c. They start out in the bone marrow and mature in the thymus.

   d. They start in lymph nodes and mature inside the thyroid.

Having a Spleen-did Time with the Lymphatic Organs

Although the lymph nodes are the most numerous lymphatic organs, several other vital organs exist in the lymphatic system, including the spleen, thymus gland, and tonsils.

Reuse and recycle: The spleen

The spleen, the largest lymphatic organ in the body, is a 5-inch, roughly egg-shaped structure to the left of and slightly behind the stomach. Like lymph nodes, it has a hilus through which the splenic artery, splenic vein, and efferent vessels pass. (The efferent vessels transport lymph away from the spleen to lymph nodes; as we note in the preceding section, remember "e" for "exit.") Also like lymph nodes, the spleen is surrounded by a fibrous capsule that folds inward to section it off. Arterioles leading into each section are surrounded by masses of developing lymphocytes that give those areas of so-called *white pulp* their appearance. On the outer edges of each compartment, tissue called *red pulp* consists of blood-filled cavities. Unlike lymph nodes, the spleen doesn't have any afferent (access) lymph vessels, which means that it doesn't filter lymph, only blood.

Blood flows slowly through the spleen to allow it to remove microorganisms, exhausted erythrocytes (red blood cells), and any foreign material that may be in the stream. Among its various functions, the spleen can be a blood reservoir. When blood circulation drops while the body is at rest, the spleen's vessels can dilate to store any excess volume. Later, during exercise or if oxygen concentrations in the blood begin to drop, the spleen's blood vessels constrict and push any stored blood back into circulation.

But the spleen's primary role is as a biological recycling unit, capturing and breaking down defective and aged red blood cells to reuse their components later. Iron stored by the spleen's macrophages goes to the bone marrow where it's turned into hemoglobin in new blood cells. By the same token, bilirubin for the liver is generated during breakdown of hemoglobin. The spleen produces red blood cells during embryonic development but shuts down that process after birth; in cases of severe anemia, the spleen sometimes starts up production of red blood cells again.

Fortunately, the spleen isn't considered a vital organ; if it is damaged or has to be surgically removed, the liver and bone marrow can pick up where the spleen leaves off.

T cell central: The thymus gland

Tucked just behind the breastbone and between the lungs in the upper chest (the *superior mediastinum,* if you want to be technical), the thymus gland was a medical mystery until recent decades. Its two oblong lobes are largest at puberty when they weigh around 40 grams (somewhat less than an adult mouse). Through a process called *involution,* however, the gland atrophies and shrinks to roughly 6 grams by the time an adult is 65. (You can remember that term as the inverse of evolution.)

The thymus gland serves its most critical role — as a nursery for immature T lymphocytes, or T cells — during fetal development and the first few years of a human's life. Prior to birth, fetal bone marrow produces *lymphoblasts* (early stage lymphocytes) that migrate to the thymus. Shortly after birth and continuing until adolescence, the thymus secretes several hormones, collectively called *thymosin,* that prompt the early cells to mature into full-grown T cells that are *immunocompetent,* ready to go forth and conquer invading microorganisms. (These hormones are the reason the thymus is considered part of the endocrine system, too; check out Chapter 16 for details on this system.)

As with other lymphatic structures, the thymus is surrounded by a fibrous capsule that dips inside to create chambers called *lobules*. Within each lobule is a cortex made of T cells held in place by reticular fibers and a central medulla of unusually onionlike layered epithelial cells called *thymic corpuscles,* or *Hassall's corpuscles,* as well as scattered lymphocytes.

Open wide and move along: The tonsils and Peyer's patches

Like the thymus gland, the tonsils, which are misunderstood masses of lymphoid tissue, are largest around puberty and tend to atrophy as an adult ages. Unlike the thymus, however, the tonsils don't secrete hormones but do produce lymphocytes and antibodies to protect against microorganisms that are inhaled or eaten. Although only two are visible on either side of the pharynx, there are actually five tonsils:

- The two you can identify, which are called *palatine tonsils*
- The *adenoid* or *pharyngeal tonsil* in the posterior wall of the nasopharynx
- The *lingual tonsils,* which are round masses of lymphatic tissue arranged in two approximately symmetrical collections that cover the posterior one-third of the tongue

Invaginations (ridges) in the tonsils form pockets called *crypts,* which trap bacteria and other foreign matter.

Peyer's patches, also called *aggregate glands* or *agminate glands,* are masses of lymph nodules just below the surface of the ileum, the lowest section of the small intestine. When harmful microorganisms get into the intestine, Peyer's patches can mobilize an army of B cells and macrophages to fight off infection.

You've absorbed a lot in this section. See how much of it is getting caught in your filters:

16. What role does the spleen have besides that of a biological recycling unit?

   a. It undergoes involution as a person ages to make way for enlarged lymph nodes.

   b. It divides the pharynx from the tonsils to ensure that lymph fluid moves in the right direction.

   c. It produces neutrophils to maintain immunocompetent T cells.

   d. It has vessels that dilate to store excess blood volume when the body is at rest.

17. Why does the thymus stop growing during adolescence and atrophy with age?

   a. It serves its most critical role during fetal development and the first few years of life.

   b. The increasing number of lymph nodes over time takes over its role in the body.

   c. The reticular fibers surrounding it contract and crush it over time.

   d. It slowly runs out of Hassall's corpuscles.

18. Lymphoid tissue(s) located in the pharynx that protect(s) against inhaled or ingested pathogens and foreign substances is/are called the

    a. thymus.

    b. tonsils.

    c. Peyer's patches.

    d. spleen.

19. Lymphatic nodules found in the ileum of the small intestines are

    a. Peyer's patches.

    b. lymph nodes.

    c. thymus.

    d. macrophages.

20. What is the significance of the spleen having no afferent vessels?

    a. Foreign materials can't get inside and cause problems.

    b. It means that the spleen filters only blood, not lymph.

    c. It's why humans can survive removal of the spleen.

    d. The spleen evolved that way to prevent phagocytosis.

21. The lymphatic organ found in the superior mediastinum is the

    a. tonsil.

    b. spleen.

    c. thymus.

    d. reticular formation.

22. Why is the spleen's white pulp that color?

    a. Blood has drained from the red pulp back into the bloodstream.

    b. Cartilage is beginning to form at the borders.

    c. Lymphocytes clump there and give it its color.

    d. The elastic connective tissue has stretched and relaxed repeatedly over the course of a person's lifetime.

23. The lymphatic organ(s) responsible for removal of aged and defective red blood cells from the bloodstream is/are the

    a. tonsils.

    b. spleen.

    c. Peyer's patches.

    d. lymph nodes.

24. Why does the thymus secrete hormones?

    a. To make T lymphocytes immunocompetent

    b. To help the tonsils make more T lymphocytes

    c. To support the Peyer's patches as they produce B lymphocytes

    d. To signal the spleen that it's time to release more red blood cells

25. Which of the following is not a true tonsil?

    a. Pharyngeal tonsil

    b. Palatine tonsils

    c. Adenoid

    d. Sublingual tonsil

26.–29. Mark the statement with a T if it's true or an F if it's false:

    26. _____ The spleen filters both lymph and blood.

    27. _____ The thymus gland is functional in the early years of life and is most active in old age.

    28. _____ Tonsils function to protect against pathogens and foreign substances that are inhaled or ingested.

    29. _____ The spleen functions in the removal of aged and defective red blood cells and platelets from the blood.

30.–32. Fill in the blanks in the following sentences.

    30. The bilobular thymus gland is located in the _____.

    31. _____ are masses of lymphatic nodules found in the distal portion of the small intestine.

    32. The largest lymphatic organ in the body is the _____.

Answers to Questions on the Lymphatic System

The following are answers to the practice questions presented in this chapter.

1 The lymphatic system plays an important role in regulating **b. interstitial fluid protein.** By keeping the interstitial fluid volume between tissue cells in balance, the lymphatic system also keeps the body in homeostasis (as we discuss in greater detail in Chapter 16).

2 Under what circumstances do the walls of lymphatic capillaries open? **b. When the pressure outside the capillaries is greater than inside.** The difference in pressure causes the filaments anchoring the thin layer of cells to open. Increasing pressure inside the capillary walls eventually forces them closed again.

3 How does lymph move around inside its vessels? **c. Slowly, through peristalsis, the action of semilunar valves, and the squeezing of surrounding skeletal muscles.** You already know the first answer is wrong because lymph nodes aren't pumps, and because you know the lymphatic system doesn't have pumps, you know the last answer is wrong because there is nothing steady or rhythmic about it. You're left with one of the "slowly" answers. The second answer can't be right because the system wouldn't have room to take in more lymph if what was already there wasn't moving through some kind of action. Voilà! You have the right answer.

4 What is the largest lymphatic vessel in the body? **c. Thoracic duct.** Yes, this duct is the largest lymphatic vessel. "Spleen" isn't the correct answer because that's the largest lymphatic *organ*.

5 The lymphatic system serves several crucial functions, but what does it *not* normally do? **d. Produce erythrocytes.** Those are red blood cells, which develop in the bone marrow.

6–7 Lymph moves from a/an **6. afferent lymphatic** vessel through a lymph node and back out into the blood through a/an **7. efferent lymphatic** vessel. Not sure which vessel is which? A comes first in the alphabet, and you can think of it as the lymph's Access point. E comes next, and it provides an Exit point.

8 What do lymph nodes use to destroy bacteria, cancer cells, and other foreign material? **d. Macrophages.** You can arrive at the correct answer through a process of elimination. Neither stroma nor trabeculae can be right because you already know that they're both part of connective tissue. Peyer's patches are limited to the small intestine. So macrophages must be the front-line warriors here.

9 Why do lymph nodes sometimes swell when an infection is present? **c. To accommodate cellular activity of the immune-system-attacking microorganisms.** You can eliminate a couple of answers right off the bat here. The first answer can't be right; why would there be a need for additional lymph fluid? You also know the word "gastric" in the second answer relates to digestion, not the lymphatic system.

10 Lymph nodes have two primary functions. What are they? **c. To produce lymphocytes and antibodies and to filter lymph fluid.** Be careful that both parts of the answer are correct.

11 What type of cell can be found in a lymph node's medulla? **d. Both a and c.** In other words, you can find both T cells and B cells.

12 The germinal center of lymph nodules produces what type of cells? **b. Lymphocytes.** Your answer hint is in the question: *Lymph* nodules produce *lymph*ocytes.

13 The connective tissue fiber that forms the framework of the lymphoid tissue is **c. reticular.** It provides both a tissue framework and a type of netting to hold clusters of lymphocytes.

14 What shape is a lymph node? **d. Oval and slightly bent, like a kidney bean.** Alas, you're going to have to memorize some things. This fact is one of them.

15 Why are T cells called by that name? **c. They start out in the bone marrow and mature in the thymus.** That's "T" for thymus, *not* "T" for thyroid. But you know T cells have no affiliation with the thyroid, right?

16 What role does the spleen have besides that of a biological recycling unit? **d. It has vessels that dilate to store excess blood volume when the body is at rest.** Enlarged lymph nodes don't travel, so you know the first answer is wrong. Plus, you know the spleen is in the abdomen, not the throat, so toss out that second answer. Wavering between the third and fourth answers? Just remember all those TV shows where internal bleeding is chalked up to a damaged spleen, and you've got it!

17 Why does the thymus stop growing during adolescence and atrophy with age? **a. It serves its most critical role during fetal development and the first few years of life.**

Here's a memory tool that only word-play students will love: "The thymus runs out of thyme."

18 Lymphoid tissue(s) located in the pharynx that protect(s) against inhaled or ingested pathogens and foreign substances is/are called the **b. tonsils.** When you remember that the pharynx is the throat, this question becomes more obvious.

19 Lymphatic nodules found in the ileum of the small intestines are **a. Peyer's patches.** It's almost like they're "patched" onto the ileum.

20 What is the significance of the spleen having no afferent vessels? **b. It means that the spleen filters only blood, not lymph.** No afferent vessels means no access for lymph to get in there.

21 The lymphatic organ found in the superior mediastinum is the **c. thymus.**

Break this question into parts and it becomes easier to locate which gland is being referenced: *Superior* means "upper," *media–* means "middle" (or "midline"), and *–stinum* refers to the sternum, or breastbone.

22 Why is the spleen's white pulp that color? **c. Lymphocytes clump there and give it its color.** Red pulp is always red, so say goodbye to that first answer. The spleen doesn't need cartilage, so toss out the second answer. And the final answer can't be right because what would be the evolutionary point in scarring up the spleen as a person ages?

23 The lymphatic organ(s) responsible for removal of aged and defective red blood cells from the bloodstream is/are the **b. spleen.** It recycles critical components from the spent red blood cells and sends them to the bone marrow to be turned into fresh cells.

24 Why does the thymus secrete hormones? **a. To make T lymphocytes immunocompetent.** Plus, it's where these cells get the "T" in their name.

25 Which of the following is not a true tonsil? **d. Sublingual tonsil.** If it's not pharyngeal, palatine, or lingual, it's not a real tonsil. (Pharyngeal tonsils and adenoids are the same thing.)

26 The spleen filters both lymph and blood. **False.**

27 The thymus gland is functional in the early years of life and is most active in old age. **False.** The opposite is true; the thymus is practically nonexistent in old age.

28 Tonsils function to protect against pathogens and foreign substances that are inhaled or ingested. **True.**

29 The spleen functions in the removal of aged and defective red blood cells and platelets from the blood. **True.**

30 The bilobular thymus gland is located in the **superior mediastinum.** Were you thrown off by the term "bilobular"? Remember: *Bi–* means "two" and *–lobular* refers to lobes.

31 **Peyer's patches** are masses of lymphatic nodules found in the distal portion of the small intestine. Don't let that "distal" fool you; just think of it as "distant."

32 The largest lymphatic organ in the body is the **spleen.** You may be tempted to write "thoracic duct" here, but that's incorrect because the duct is the largest vessel, not the largest organ.

Chapter 12

Filtering Out the Junk: The Urinary System

*I*f you read Chapter 9 on the digestive system, you may be chewing on the idea that undigested food is the body's primary waste product. But it's not — that title belongs to urine. We make more of it than we do feces — in fact, our bodies are making small amounts of urine all the time — and we release it more often throughout the day. Most important, urine captures all the leftovers from our cells' metabolic activities and jettisons them before they can build up and become toxic. In addition, urine helps maintain *homeostasis,* or the proper balance of body fluids: electrolytes that control acid-base ratio, and the mixture of salt and water that regulates blood pressure.

In short, the urinary system

✔ Excretes useless and harmful material that it filters from blood plasma, including urea, uric acid, creatinine, and various salts

✔ Removes excess materials, particularly anything normally present in the blood that builds up to excessive levels

✔ Maintains proper osmotic pressure, or fluid balance, by eliminating excess water when concentration rises too high at the tissue level

✔ Reabsorbs water, glucose, and amino acids

✔ Produces the hormones calcitriol (the hormonally active form of vitamin D that increases blood calcium levels) and erythropoietin (which stimulates red blood cell production in the bone marrow), as well as the enzyme renin (which regulates the body's mean arterial blood pressure by controlling the extracellular fluid volume and arterial vasoconstriction; also known as *angiotensinogenase*).

This chapter explains how the urinary system collects, manages, and excretes the waste that the body's cells produce as they go about busily metabolizing all day. You discover the parts of the kidneys, ureter, urinary bladder, and urethra.

Examining the Kidneys, the Body's Filters

The kidneys are nonstop filters that sift through 1.2 liters of blood per minute. Humans have a pair of kidneys just above the waist (lumbar region) toward the back of the abdominal cavity. Although sometimes the same size, the left kidney tends to be a bit larger than the right. The last two pairs of ribs surround and protect each kidney, and a layer of fat, called *perirenal (perinephric) fat,* provides additional cushioning. Kidneys are *retroperitoneal,* which means that they're posterior to the peritoneum. The *renal capsule,* or outer lining of the kidney, is a layer of collagen fibers; these fibers extend outward to anchor the organ to surrounding structures.

Each kidney is dark red, about 4½ inches long, and shaped like a bean (hence the type of legumes called kidney beans). The portion of the bean that folds in on itself, referred to as the *medial border,* is concave with a deep depression in it called the *hilus,* or *hilum.* The hilus opens into a fat-filled space called the *renal sinus,* which in turn contains the *renal pelvis, renal calyces,* blood vessels, nerves, and fat. The *renal artery* and *renal vein,* which provide the kidney's blood supply, as well as the ureter that carries urine to the bladder, leave the kidney through the hilus.

Immediately below the renal capsule is a granular layer called the *renal cortex,* and just below that is an inner layer called the *medulla* that folds into anywhere from 8 to 18 conical projections called the *renal pyramids.* Between the pyramids are *renal columns* that extend from the cortex inward to the renal sinus. The tips of these pyramids, the *renal papillae,* empty their contents into a collecting area called the *minor calyx.* It's one of several saclike structures referred to as the *minor* and *major calyces* which form the start of the urinary tract's "plumbing" system and collect urine transmitted through the papillae from the cortex and medulla. Although the number varies between individuals, generally each of two or three major calyces branches into four or five minor calyces, with a single minor calyx surrounding the papilla of one pyramid. Urine passes through the minor calyx into its major calyx and then into the ureter for the trip to the bladder.

The following sections dig deeper into the kidneys' microscopic structure and filtration process.

Going microscopic

At the microscopic level, each kidney contains more than one million tiny tubes known as *uriniferous tubules,* or *nephrons.* These are the primary functional units of the urinary system. At one end, each nephron is closed off and folded into a small double-cupped structure called a *Bowman's capsule,* or the *glomerular capsule,* where the actual process of filtration occurs. Leading away from the capsule, the nephron forms into the *first,* or *proximal, convoluted tubule* (PCT), which is lined with cuboidal epithelial cells having microvilli brush borders that increase the area of absorption. This tube straightens to form a structure called the *descending loop of Henle* and then bends back in a hairpin turn into another structure called the *ascending loop of Henle.* After that, the tube becomes convoluted again, forming the *second,* or *distal, convoluted tubule* (DCT), which is made of the same types of cells as the first, or proximal, convoluted tubule but without any microvilli. This tubule connects to a collecting tubule that it shares with the output ends of many other nephrons. The collecting tubules open into the minor calyces of the renal pelvis, which in turn open into the major calyces.

Because of their role as the body's key filters, the kidneys receive about 20 percent of all the blood pumped by the heart each minute. A large branch of the abdominal aorta, called the *renal artery,* carries that blood to the kidneys. After branching into smaller and smaller vessels, the blood eventually enters interlobular arteries to the *afferent arterioles,* each of which branches into tufts of five to eight capillaries called a *glomerulus* (the plural is *glomeruli*) inside the glomerular (Bowman's) capsule. In the glomerulus, pressure differences force filtration of solutes, fluids, and other glomerular filtrates through the capillary walls into the glomerular (Bowman's) capsule. The glomerular capillaries come back together to form *efferent arterioles,* which then branch to form the *peritubular,* or *second, capillary bed* surrounding the convoluted tubules, the loop of Henle, and the collecting tubule. The capillaries come together once again to form a small vein *(venule)* that empties blood into the interlobular vein, eventually moving the blood into the renal vein to depart the kidneys.

Each glomerulus and its surrounding glomerular (Bowman's) capsule make up a single *renal corpuscle* where basic filtration takes place. Like all capillaries, glomeruli have thin, membranous walls, but unlike their capillary cousins elsewhere, these vessels have unusually large pores called *fenestrations* or *fenestrae* (from the Latin word *fenestra* for "window").

Focusing on filtering

To understand how the renal corpuscles work, think of an espresso machine: Water is forced under pressure through a sieve containing ground coffee beans, and a filtrate called brewed coffee trickles out the other end. Something similar takes place in the renal corpuscles. Hydrostatic pressure forces fluids across the glomerular membranes, which capture about 125 milliliters of material per minute in the glomerular (Bowman's) capsules. The glomerular filtration membrane is thousands of times more permeable to water and solutes than are other capillaries. The glomerular blood pressure is much higher than that of other capillary beds: approximately 55 to 60 millimeters of mercury (mm Hg), as opposed to 15 to 20 millimeters of mercury in other capillary beds. Selective reabsorption occurs mostly in the PCT as these filtered materials then move through the nephrons' network of tubules, returning the bulk of the water and much of the needed materials back to the bloodstream through the peritubular (secondary) capillary bed surrounding the nephrons' structures. Tubular secretions also occur, which transfer material from the peritubular capillaries to the renal tubules through active transport. Only a few substances are secreted back into the renal tubules. These substances are present in great excess or are poisons.

So, despite 125 milliliters of material coming out of the blood every minute, only 1 milliliter of urine is generated each minute. This is a matter of simple subtraction: Reabsorption of about 100 milliliters per minute takes place in the proximal convoluted tubules. The loop of Henle returns 7 milliliters per minute more. The distal convoluted tubules return 12 milliliters, and the collecting tubules return about 5 milliliters. Voilà! That totals 124 milliliters of reabsorption per minute and explains the 1 milliliter of urine that comes out when all is said and done.

While all this filtering and absorption is going on, the kidneys also sometimes secrete an enzyme called *renin* (also known by its more complicated chemical name of *angiotensinogenase*) that converts a peptide generated in the liver, called *angiotensinogen,* into *angiotensin I.* Angiotensin I moves into the lungs where a converting enzyme turns it into *angiotensin II,* a potent *vasoconstrictor.* Say what? Try this explanation, instead: The kidneys work to ensure that systemic blood pressure remains high enough for them to do their filtering job properly. That's what a *vasoconstrictor* is: a substance that causes blood vessels to narrow, increasing the pressure of the fluids moving through them. Rising blood pressure also triggers the adrenal glands perched atop each kidney to release *aldosterone,* causing the renal tubules to absorb more sodium with water following it that pumps up blood volume. The pituitary gland also plays a role in urine production by releasing an antidiuretic hormone (ADH) that causes water retention at the kidneys and elevated blood pressure.

You've absorbed a lot in the last few paragraphs. See how much of it is getting caught in your filters.

1.–6. Use the following terms to identify the parts of the kidney in Figure 12-1.

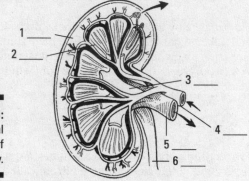

Figure 12-1:
Internal anatomy of the kidney.

Illustration by Kathryn Born, MA

a. Renal vein

b. Cortex

c. Ureter

d. Renal pelvis

e. Renal artery

f. Medulla (renal pyramids)

7.–14. Use the following terms to identify the nephron structures in Figure 12-2.

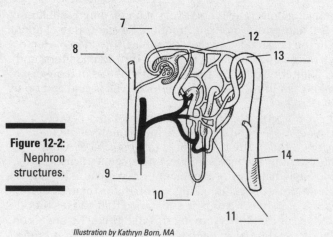

Figure 12-2:
Nephron structures.

Illustration by Kathryn Born, MA

a. Loop of Henle

b. Glomerular (Bowman's) capsule

c. Collecting tubule

d. Proximal convoluted tubule

e. Interlobular vein

f. Peritubular (secondary) capillary bed

g. Distal convoluted tubule

h. Interlobular artery

15.–19. Match the anatomical terms with their descriptions.

15. _____ Cortex

16. _____ Medulla

17. _____ Renal pelvis

18. _____ Calyx

19. _____ Collecting tubule

a. Composed of folds forming the renal pyramids

b. Granular outer layer

c. Irregular, saclike structures for collecting urine in the renal pelvis

d. Transports urine from the nephrons

e. Found in the renal sinus

20. The human kidney lies outside the abdominal cavity. This makes it

a. retroperitoneal.

b. parietal.

c. endocoelomic.

d. exterocoelomic.

21. What is the primary functional unit of a kidney?

a. The medulla

b. The outer layers covering Henle's loop

c. The renal papillae

d. Uriniferous tubules, or nephrons

22. The correct sequence for removal of material from the blood through the nephron is

a. afferent arteriole → glomerulus → proximal convoluted tubule → loop of Henle → distal convoluted tubule → collecting tubule.

b. afferent arteriole → glomerulus → distal convoluted tubule → loop of Henle → proximal convoluted tubule → collecting tubule.

c. afferent arteriole → collecting tubule → glomerulus → proximal convoluted tubule → loop of Henle → distal convoluted tubule.

d. efferent arteriole → proximal convoluted tubule → glomerulus → loop of Henle → distal convoluted tubule → collecting tubule.

23. Where are brush borders of villi found primarily?

    a. Ascending loop of Henle

    b. Proximal convoluted tubule

    c. Glomerular (Bowman's) capsule

    d. Descending loop of Henle

24. Although the kidney has many functional parts, where does filtration primarily occur?

    a. Inside the distal convoluted tubules

    b. Outside the loop of Henle

    c. Just beyond the renal cortex

    d. Within the glomerulus inside a glomerular (Bowman's) capsule

25. If 125 milliliters of material are captured per minute in the glomerular (Bowman's) capsules, why is only 1 milliliter of urine generated each minute?

    a. The renal pelvis siphons off most of it.

    b. A series of tubules allow most of it to be reabsorbed into the bloodstream.

    c. The material is highly concentrated by the time it becomes urine.

    d. Hydrostatic pressure forces most of the material back out again.

Getting Rid of the Waste

After your kidneys filter out the junk, it's time to deliver it to the bladder and the urethra via the ureters. The following sections explain how that's done.

Surfing the ureters

Ureters are narrow, muscular tubes through which the collected waste travels. About 10 inches long, each ureter descends from a kidney to the posterior lower third of the bladder. Like the kidneys themselves, the ureters are behind the peritoneum outside the abdominal cavity, so the term *retroperitoneal* applies to them, too. The inner wall of the ureter is a simple mucous membrane composed of elastic transitional epithelium. It also has a middle layer of smooth muscle tissue that propels the urine by peristalsis — the same process that moves food through the digestive system. So rather than trickling into the bladder, urine arrives in small spurts as the muscular contractions force it down. The tube is surrounded by an outer fibrous layer of connective tissue that supports it during peristalsis.

Ballooning the bladder

The urinary *bladder* is a large, muscular bag that lies in the pelvis behind the pubis bones. In human females, it's beneath the uterus and in front of the vagina (see Chapter 14). In human males, the bladder lies between the rectum and the symphysis pubis (see Chapter 13). There are three openings in the bladder: two on the back side where the ureters enter and one on the front for the *urethra,* the tube that carries urine outside the body (see the next section). The bladder's *trigone* is the triangular area between these three openings. The *neck* of the bladder surrounds the urethral attachment, and the *internal sphincter* (smooth muscle that provides involuntary control) encircles the junction between the urethra and the bladder.

Inside, the bladder is lined with highly elastic transitional epithelium tissue (described in Chapter 4). When full, the bladder's lining is smooth and stretched; when empty, the lining lies in a series of folds called *rugae* (just as the stomach does). When the bladder fills, the increased pressure stimulates the organ's stretch receptors, prompting the individual to urinate. The bladder can hold 600–800 milliliters of urine, but usually the pressure receptor's response will cause it to empty before reaching maximum capacity. (See the later section "Spelling relief: Urination" for more details.)

Distinguishing the male and female urethras

Both males and females have a *urethra,* the tube that carries urine from the bladder to a body opening, or orifice. Both males and females have an internal sphincter controlled by the autonomic nervous system and composed of smooth muscle tissue to guard the exit from the bladder. Both males and females also have an external sphincter composed of circular striated, or skeletal, muscle tissue that's under voluntary control. But as we all well know, the exterior plumbing is rather different.

The female urethra is about one and a half inches long and lies close to the vagina's anterior (front) wall. It opens just in front of the vaginal opening. The external sphincter for the female urethra lies just inside the urethra's exit point.

The male urethra is about 8 inches long and carries a different name as it passes through each of three regions:

- The *prostatic urethra* leaving the bladder contains the internal sphincter and passes through the prostate gland. Several openings appear in this region of the urethra, including a small opening where sperm from the vas deferens and ejaculatory duct enters, and prostatic ducts where fluid from the prostate enters.

- The *membranous urethra* is a small 1- or 2-centimeter portion that contains the external sphincter and penetrates the pelvic floor. The tiny *bulbourethral* (or *Cowper's) glands* lie on either side of this region.

- The *cavernous urethra,* also known as the *spongy,* or *penile, urethra,* runs the length of the penis on its ventral surface through the corpus spongiosum, ending at a vertical slit at the end of the penis. Ducts from the Cowper's *(bulbourethral)* glands enter at this region.

Spelling relief: Urination

Urination, known by the medical term *micturition,* occurs when the bladder is emptied through the urethra. Although urine is created continuously, it's stored in the bladder until the individual finds a convenient time to release it. Mucus produced in the bladder's lining protects its walls from any acidic or alkaline effects of the stored urine. When there is about 200 milliliters of urine distending the bladder walls, stretch receptors transmit impulses to warn that the bladder is filling. *Afferent* impulses are transmitted to the spinal cord, and *efferent* impulses return to the bladder, forming a reflex arc that causes the internal sphincter to relax and the muscular layer of the bladder to contract, forcing urine into the urethra. The afferent impulses continue up the spinal cord to the brain, creating the urge to urinate. Because the external sphincter is composed of skeletal muscle tissue, no urine usually is released until the individual voluntarily opens the sphincter.

Now test your knowledge of how the human body gets rid of its waste:

Q. The separation of the reproductive and urinary systems is complete in the human

   a. male.

   b. female.

   c. male and female.

A. The correct answer is female. The male urethra runs through the same "plumbing" as the male reproductive system.

26. How can the interior of the bladder expand as much as it does?

   a. Its ciliated columnar epithelium can shift away from each other.

   b. It is lined with transitional epithelium.

   c. Its cuboidal epithelium can realign its structure as needed.

   d. Its white fibrous connective tissue gives way under pressure.

27. How does urine move from the kidney down the ureter?

   a. By fibrillation

   b. By flexure

   c. Simply because of gravity

   d. By peristaltic contractions

28. The external sphincter of the urinary tract is made up of

   a. circular striated (skeletal) muscle.

   b. rugae.

   c. simple mucous membrane.

   d. the membranous urethra.

29. What is the function of the internal sphincter at the junction of the bladder neck and the urethra?

   a. To stimulate the expulsion of urine from the bladder

   b. To keep foreign substances and infections from entering the bladder

   c. To prevent urine leakage from the bladder

   d. To prevent return of urine from the urethra back into the bladder

30. True or false: Urine production occurs sporadically throughout the day.

Answers to Questions on the Urinary System

The following are answers to the practice questions presented in this chapter.

1–6 Following is how Figure 12-1, the internal anatomy of the kidney, should be labeled.

1. **b. Cortex;** 2. **f. Medulla (renal pyramids);** 3. **d. Renal pelvis;** 4. **e. Renal artery;** 5. **a. Renal vein;** 6. **c. Ureter**

7–14 Following is how Figure 12-2, the nephron, should be labeled.

7. **b. Glomerular (Bowman's) capsule;** 8. **h. Interlobular artery;** 9. **e. Interlobular vein;** 10. **a. Loop of Henle;** 11. **f. Peritubular (secondary) capillary bed;** 12. **d. Proximal convoluted tubule;** 13. **g. Distal convoluted tubule;** 14. **c. Collecting tubule**

15 Cortex: **b. Granular outer layer.**

16 Medulla: **a. Composed of folds forming the renal pyramids.**

17 Renal pelvis: **e. Found in the renal sinus.**

18 Calyx: **c. Irregular, saclike structures for collecting urine in the renal pelvis.**

19 Collecting tubule: **d. Transports urine from the nephrons.**

20 The human kidney lies outside the abdominal cavity. This makes it **a. retroperitoneal.** *Peritoneal* refers to the peritoneum, the membrane lining the abdominal cavity; and *retro* can be defined as "situated behind."

21 What is the primary functional unit of a kidney? **d. Uriniferous tubules, or nephrons.** Each nephron contains a series of the parts needed to do the kidney's filtering job.

22 The correct sequence for removal of material from the blood through the nephron is **a. afferent arteriole → glomerulus → proximal convoluted tubule → loop of Henle → distal convoluted tubule → collecting tubule.** In short, blood comes through the artery (arteriole) and material gloms onto the nephron before twisting through the near (proximal) tubes, looping the loop, twisting through the distant (distal) tubes, and collecting itself at the other end. Try remembering artery-glom-proxy-loop-distant-collect.

23 Where are brush borders of villi found primarily? **b. Proximal convoluted tubule.** Those brush borders provide extra surface area for reabsorption, so it makes sense that they congregate in the first area after filtration.

24 Although the kidney has many functional parts, where does filtration primarily occur? **d. Within the glomerulus inside a glomerular (Bowman's) capsule.** The glomerulus is a collection of capillaries with big pores, so think of it as the initial filtering sieve.

25 If 125 milliliters of material are captured per minute in the glomerular (Bowman's) capsules, why is only 1 milliliter of urine generated each minute? **b. A series of tubules allow most of it to be reabsorbed into the bloodstream.** Remember the subtraction puzzle we discuss earlier in this chapter? We count up 124 milliliters of reabsorption.

26 How can the interior of the bladder expand as much as it does? **b. It is lined with transitional epithelium.** It's transitional because it needs to be able to stretch and collapse as needed.

27 How does the urine move from the kidney down the ureter? **e. By peristaltic contractions.** It's the same action that moves food through the digestive system.

28 The external sphincter of the urinary tract is made up of **a. circular striated (skeletal) muscle.** The external sphincter is under a person's control. The only tissue in this list of choices that features voluntary control is skeletal muscle.

29 What is the function of the internal sphincter at the junction of the bladder neck and the urethra? **c. To prevent urine leakage from the bladder.** The sphincter's makeup of smooth muscle tissue helps it do so.

30 Urine production occurs sporadically throughout the day. **False.** The kidneys produce urine continuously in a healthy individual.

Part IV
Survival of the Species

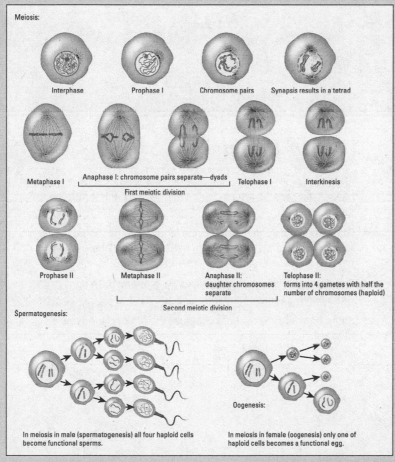

Illustration by Imagineering Media Services Inc.

In this part . . .

✔ Identify the parts of the male reproductive system and explore how chromosomes are packaged to prepare them for their journey.

✔ Follow the female menstrual cycle to understand the nuances of fertility and discover how one cell becomes 32 before the embryo takes up residence in the uterus.

✔ Track the progress from embryo to fetus to newborn baby (technically called a *neonate*).

✔ Jump from infancy to senescence as the life cycle moves forward.

Chapter 13

Why Ask Y? The Male Reproductive System

*I*ndividually, humans don't need to reproduce to survive. But to survive as a species, a number of individuals must produce and nurture a next generation, carrying their uniqueness forward in the genetic pool. Humans are born with the necessary organs to do just that.

In this chapter, you get an overview of the parts and functions of the male reproductive system, along with plenty of practice questions to test your knowledge. (We cover the guys first because their role in the basic reproduction equation isn't nearly as long or complex as that of their mates. We address the female reproductive system in Chapter 14.)

Identifying the Parts of the Male Reproductive System

On the outside, the male reproductive parts, which you can see in Figure 13-1, are straight-forward — a *penis* and a *scrotum.* At birth, the apex of the penis is enclosed in a fold of skin called the *prepuce,* or *foreskin,* which often is removed during a surgery called *circumcision.*

The scrotum is a pouch of skin divided in half on the surface by a ridge called a *raphe* that continues up along the underside of the penis and down all the way to the anus. The left side of the scrotum tends to hang lower than the right side to accommodate a longer *spermatic cord,* which we explain later in this section.

There are two scrotal layers: the *integument,* or outer skin layer, and the *dartos tunic,* an inner smooth muscle layer that contracts when cold and elongates when warm. Why? That has to do with the two *testes* (the singular is *testis*) inside (see Figure 13-2). These small ovoid glands, also referred to as *testicles,* need to be a bit cooler than body temperature in order to produce viable sperm for reproduction. When the dartos tunic becomes cold, such as when a man is swimming, it contracts and draws the testes toward the body for warmth. When the dartos tunic becomes overly warm, it elongates to allow the testes to hang farther away from the heat of the body.

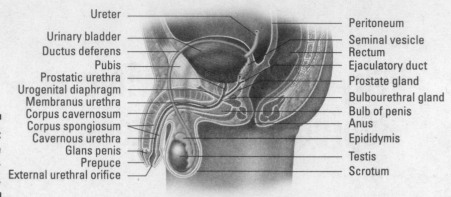

Figure 13-1:
The male reproductive system.

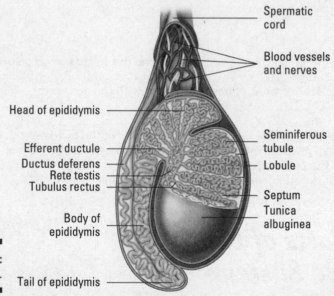

Figure 13-2:
Testis.

A fibrous capsule called the *tunica albuginea* encases each testis and extends into the gland forming incomplete *septa* (partitions), which divide the testis into about 200 *lobules*. These compartments contain small, coiled *seminiferous tubules. Spermatogenesis,* or *meiosis,* takes place in the seminiferous tubules containing cells in various stages. Spermatogenic (sperm-forming) cells are produced by a series of cellular divisions. These cells, called *spermatogonia,* divide continually by mitosis until puberty. (You can find out more about mitosis in Chapter 3.) Spermatogenesis begins during puberty when a spermatogonium divides mitotically to produce two types of daughter cells:

- ✔ **Type A cells** remain in the tubule, producing more spermatogonia.

- ✔ **Type B cells** produce four sperm in a meiosis process that we review in the next section.

Cells of the seminiferous tubules, called *sustentacular cells* (also known as *Sertoli cells* or *nurse cells*), are activated by follicle stimulating hormone (FSH), which is synthesized and secreted by the gonadotrophes of the pituitary gland. (See Chapter 16 for more information on these topics.) The nurse cells secrete androgen-binding protein that concentrates the male sex hormone *testosterone* close to the germ cells to maintain the environment the spermatogonia need to develop and mature.

Distributed in gaps between the tubules are interstitial cells called *Leydig cells* that produce testosterone. The tubules of each lobule come together in an area called the *mediastinum testis* and straighten into *tubuli recti* before forming a network called the *rete testis* that leads to the *efferent ducts* (also called *ductules*). These ducts carry sperm to an extremely long (about 20 feet), tightly coiled tube called the *epididymis* for storage and maturation of the sperm.

The epididymis merges with the *ductus deferens,* or *vas deferens,* which carries sperm up into the spermatic cord, which also encases the testicular artery and vein, lymphatic vessels, and nerves. Convoluted pouches called *seminal vesicles* lie behind the base of the bladder and secrete an alkaline fluid containing fructose, vitamins, amino acids, and prostaglandins to nourish sperm as it enters the *ejaculatory duct* (refer to Figure 13-1).

From there the mixture containing sperm enters the *prostatic urethra* that's surrounded by the *prostate gland.* This gland secretes a thin, opalescent substance that precedes the sperm in an ejaculation, contributing approximately 30 percent of the semen. The alkaline nature of this substance effectively neutralizes the acidity found in the male urethra and reduces the natural acidity of the female's vagina to prepare it to receive the sperm.

Two yellowish pea-sized bodies called *Cowper's glands,* or *bulbourethral glands,* lie on either side of the urethra and secrete a clear alkaline lubricant prior to ejaculation; it neutralizes the acidity of the urethra and acts as a lubricant for the penis. Once all the glands have added protective and nourishing fluids to the 400 to 500 million departing sperm, the mixture is known as *seminal fluid* or *semen.*

During sexual arousal, physical and/or mental stimulation causes the brain to send chemical messages via the central nervous system to nerves in the penis telling its blood vessels to relax so that blood can flow freely into the penis. A *parasympathetic reflex* is triggered that stimulates arterioles to dilate in three columns of spongy erectile tissue in the penis — the *corpus spongiosum* surrounding the urethra and the two *corpora cavernosa,* on the dorsal surface of the erect penis. As the arterioles dilate, blood flow increases while vascular shunts constrict and reroute the blood supply from the bottom to the topside of the penis. Swelling with blood, the penis further compresses its vascular drainage, resulting in a rigid erection that makes the penis capable of entering the female's vagina. Most of the rigidity of the erect penis results from blood pressure in the corpora cavernosa; high blood pressure in the corpus spongiosum could close the urethra, preventing passage of the semen.

When the male has achieved a sufficient level of stimulation, ejaculation begins. At the time of ejaculation, under the control of the *sympathetic nervous system,* sperm is expelled; rhythmic contractions of smooth muscles in the wall of the epididymis force sperm through the ductus deferens, located in the *inguinal canal,* toward the urinary bladder. After mixing with the secretions from the seminal vesicles and the prostate gland, the semen travels along the urethra and out a vertical slit in the *glans penis,* or head of the penis.

See how familiar you are with the male anatomy by tackling these practice questions:

1. Why does the left side of the scrotum tend to hang lower than the right?

   a. To adapt to changes occurring during puberty

   b. Because the left side contains the larger testicle

   c. To accommodate a longer spermatic cord

   d. Because most men are right-handed

2. The _____ is/are the source of the action when the scrotum adjusts to surrounding temperatures.

   a. testes

   b. bulbourethral (Cowper's) glands

   c. dartos tunic

   d. prostatic urethra

3.–8. Fill in the blanks in the following sentences.

Sperm-forming cells, or **3.** _____, divide throughout the reproductive lifetime of the male by a process called **4.** _____. But after puberty, these cells begin to produce two types of **5.** _____ cells. Type A cells remain in the **6.** _____, producing more of themselves. Type B cells produce **7.** _____ through a life-long process called **8.** _____.

9. Where is testosterone produced?

   a. Seminiferous tubules

   b. Interstitial (Leydig) cells

   c. Adenohypophysis

   d. Subtentacular (Sertoli or nurse) cells

10. Select the correct sequence for the movement of sperm:

   a. Seminiferous tubules → tubuli recti → rete testis → efferent ducts → epididymis → ductus deferens → ejaculatory duct → urethra

   b. Seminiferous tubules → tubuli recti → rete testis → ejaculatory duct → epididymis → ductus deferens → efferent ducts → urethra

   c. Epididymis → ejaculatory duct → tubuli recti → rete testis → efferent ducts → seminiferous tubules → ductus deferens → urethra

   d. Seminiferous tubules → ejaculatory duct → tubuli recti → rete testis → epididymis → efferent ducts → ductus deferens → urethra

11. As the sperm moves through the reproductive ducts, which of the following does not add a secretion to it?

    a. Leydig (interstitial) cells

    b. Cowper's (bulbourethral) glands

    c. Seminal vesicle

    d. Prostate

12. Sperm storage takes place in the convoluted tube called the

    a. seminiferous tubule.

    b. rete testis.

    c. spermatic cord.

    d. epididymis.

13. After all the proper glands have secreted fluids to nourish and protect the departing sperm, the substance ejaculated is called

    a. stroma.

    b. semen.

    c. prepuce.

    d. inguinal.

14. An average ejaculation will contain about _____ sperm.

    a. 40 to 50 million

    b. 400 to 500 million

    c. 400 to 500

    d. 4 to 5 million

15. During sexual arousal, the penis becomes hard and erect. What is happening to cause that?

    a. Muscles in the base of the penis are contracting while other muscles toward the end of the penis are expanding.

    b. The corpus spongiosum and the corpora cavernosa draw slowly away from one another.

    c. Blood vessels are dilating to allow free flow of blood into the penis, while vascular shunts constrict drainage and cause blood pressure to rise in the corpora cavernosa.

    d. The Cowper's glands are squeezing additional fluids into the Leydig cells.

Packaging the Chromosomes for Delivery

Sperm, the male sex cell, is produced during a process called *meiosis* (which also produces the female sex cell, or *ovum*). Meiosis involves two divisions:

- ✔ **Reduction division:** The first, a *reduction division,* divides a single *diploid* cell with two sets of chromosomes into two *haploid* cells with only one set each.

- ✔ **Division by mitosis:** The second process is a division by *mitosis* that divides the two haploid cells into four cells with a single set of chromosomes each.

Review the reproduction terms in Table 13-1.

Table 13-1	Reproduction Terms to Know		
Terms That Vary by Gender	**General Term**	**Male Term**	**Female Term**
Sex organs	Gonads	Testes	Ovaries
Original cell	Gametocyte	Spermatocyte	Oocyte
Meiosis	Gametogenesis	Spermatogenesis	Oogenesis
Sex cell	Gamete	Spermatozoan (Sperm)	Ovum

A *diploid cell* (or 2N) has two sets of chromosomes, whereas a *haploid cell* (or 1N) has one set of chromosomes.

Meiosis, which you can see in Figure 13-3, is a continuous process. Once it starts, it doesn't stop until gametes are formed. Meiosis is described in a series of stages as follows (for more on the terminology of cells, see Chapter 2):

1. **Interphase:** The original diploid cell — called a *spermatocyte* in a man and an *oocyte* in a woman — is said to be in a "resting stage," but it actually undergoes constant activity. Just before it starts to divide, the DNA molecules in the *chromanemata* (chromatin network) duplicate.

2. **Prophase I:** Structures disappear from the nucleus, including the nuclear membrane, nucleoplasm, and nucleoli. The cell's centrosome duplicates, forming two centrosomes each containing two *centrioles* that move to the ends of the nucleus and form poles. Structures begin to appear in the nuclear region, including *spindles* (protein filaments that extend between the poles) and *asters*, or *astral rays* (protein filaments that extend from the poles into the cytoplasm). The chromanemata contract, forming chromosomes. Those chromosomes then start to divide into two *chromatids* but remain attached by the centromere. *Homologous* chromosomes that contain the same genetic information pair up and go into *synapsis,* twisting around each other to form a *tetrad* of four chromatids. These tetrads begin to migrate toward the *equatorial plane* (an imaginary line between the poles).

3. **Metaphase I:** The tetrads align on the equatorial plane, attaching to the spindles by the centromere.

4. **Anaphase I:** Homologous chromosomes separate into dyads by moving along the spindles to opposite poles. In late anaphase, a slight furrowing is apparent in the cytoplasm, initiating *cytokinesis* (the division of the cytoplasm).

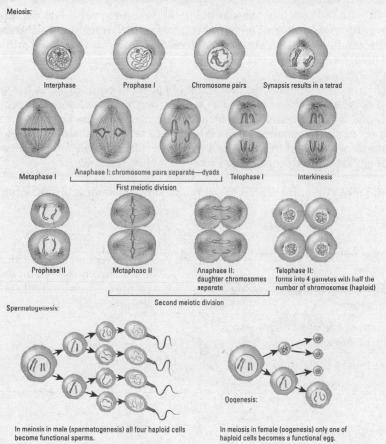

Figure 13-3:
The stages
of meiosis.

Illustration by Imagineering Media Services Inc.

5. **Telophase I:** The contracted and divided homologous chromosomes are at opposite poles. Spindle and aster structures disappear, and a nuclear membrane and nucleoplasm begin to appear in each newly forming cell. Chromosomes remain as chromatids, still contracted and divided. The furrowing seen in anaphase I continues to deepen, dividing the cytoplasm. In the male, the cytoplasm divides equally between the two cells. In the female, cytoplasmic division is unequal.

6. **Interkinesis:** The cytoplasm separates. Two genetically identical haploid cells are formed with half the number of chromosomes as the original cell. In the male, the cells are of equal size. In the female, one cell is large and the other is small.

7. **Prophase II:** The cells enter the second phase of meiosis. Once again, structures disappear from the nuclei and poles appear at the ends. Spindles and asters appear in the nuclear region. Chromosomes are already contracted and divided into chromatids attached by the centromere, and they begin to migrate toward the equatorial plane.

8. **Metaphase II:** The chromatids align on the equatorial plane and attach to the spindles by the centromere.

9. **Anaphase II:** The centromere splits as the chromatids separate, becoming chromosomes that move along the spindles to the poles. A slight furrowing appears in the cytoplasm.

10. **Telophase II:** With the chromosomes at the poles, spindles and asters disappear while new nuclear structures appear. The chromosomes uncoil, returning to chromanemata and their chromatin network. Cytoplasmic division continues to deepen and each haploid cell divides, forming four cells.

At the end of this process, the male has four haploid sperm of equal size. As the sperm mature further, a *flagellum* (tail) develops. The female, on the other hand, has produced one large cell, the secondary oocyte, and three small cells called *polar bodies;* all four structures contain just one set of chromosomes. The polar bodies eventually disintegrate and the secondary oocyte becomes the functional cell. When fertilized by the sperm, the resulting *zygote* (fertilized egg) is diploid, containing two sets of chromosomes.

Think you've conquered this process? Find out by tackling these practice questions:

Q. The metaphase II stage in meiosis involves

    a. the slipping of the centromere along the chromosome.

    b. the alignment of the chromosomes on the equatorial plane.

    c. the contraction of the chromosomes.

    d. the disappearance of the nuclear membrane.

A. The correct answer is the alignment of the chromosomes on the equatorial plane. Think "divide and conquer."

16. What happens during meiosis in the male?

    a. The gonads are differentiated by chromosome count.

    b. The first division divides a diploid cell into two haploid cells, which then split again to form four sperm.

    c. The tetrads align on the equatorial plane, setting up fresh spindles for the centromere.

    d. The dyads align at the poles and then divide to differentiate the chromosome count.

17. A man has 46 chromosomes in a spermatocyte. How many chromosomes are in each sperm?

   a. 23 pairs

   b. 23

   c. 184

   d. 46

18. Synapsis, or side-by-side pairing, of homologous chromosomes

   a. occurs in mitosis.

   b. completes fertilization.

   c. occurs in meiosis.

   d. signifies the end of prophase of the second meiotic division.

19. Anaphase I of meiosis is characterized by which of the following?

   a. Synapsed chromosomes move away from the poles.

   b. DNA duplicates itself.

   c. Synapsis of homologous chromosomes occurs.

   d. Homologous chromosomes separate and move poleward with centromeres intact.

20. During oogenesis, the three nonfunctional cells produced are called

   a. cross-over gametes.

   b. male sex cells.

   c. polar bodies.

   d. somatic cells.

21. What happens during prophase II, or the second phase of meiosis?

   a. The contracted and divided homologous chromosomes remain at opposite poles.

   b. The chromatids align on the equatorial plane and attach to the spindles by the centromere.

   c. The tetrads align on the equatorial plane, attaching to the spindles by the centromere.

   d. Structures disappear from the nuclei, and spindles and asters appear in the nuclear region.

22. Complete the following worksheet on the stages of meiosis.

Draw the stages of meiosis: Describe the changes in each stage:

1.

Late Prophase I

2.

Metaphase I

3.

Anaphase I

4.

Telophase I

5.

Interkinesis

6.

Prophase II

7.

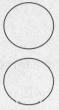

Metaphase II

8.

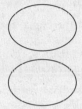

Anaphase II

9.

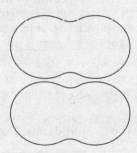

Telophase II

10.

Sperm

11.

Ovum and Polar Bodies

Answers to Questions on the Male Reproductive System

The following are answers to the practice questions presented in this chapter.

1 Why does the left side of the scrotum tend to hang lower than the right? **c. To accommodate a longer spermatic cord.** The key words here are "tend to." It's not true of all men, but in general the left testicle hangs lower than the right because it has a longer cord.

2 The **c. dartos tunic** is the source of the action when the scrotum adjusts to surrounding temperatures. This is the inner smooth muscle layer of the scrotum.

3-8 Sperm-forming cells, or **3. spermatogonia,** divide throughout the reproductive lifetime of the male by a process called **4. mitosis.** But after puberty, these cells begin to produce two types of **5. daughter** cells. Type A cells remain in the **6. seminiferous tubules,** producing more of themselves. Type B cells produce **7. four sperm** through a life-long process called **8. meiosis.**

9 Where is testosterone produced? **b. Leydig (interstitial) cells.** The word *interstitial* can be translated as "placed between," which is where Leydig cells are.

10 Select the correct sequence for the movement of sperm: **a. Seminiferous tubules → tubuli recti → rete testis → efferent ducts → epididymis → ductus deferens → ejaculatory duct → urethra.** The sperm develop in the coiled tubules, move through the straighter tubes (tubuli recti), continue across the network of the testis (rete testis), through the efferent ducts, into the epididymis (remember the really long tube), and past the ductus (or vas) deferens and the ejaculatory duct into the urethra.

11 As the sperm moves through the reproductive ducts, which of the following does not add a secretion to it? **a. Leydig (interstitial) cells.** Interstitial cells secrete testosterone, the primary male hormone.

12 Sperm storage takes place in the convoluted tube called the **d. epididymis.** The other answer options don't come into play until it's time to release semen.

13 After all the proper glands have secreted fluids to nourish and protect the departing sperm, the substance ejaculated is called **b. semen.**

14 An average ejaculation will contain about **b. 400 to 500 million** sperm. Keep in mind that sperm are microscopically small, so quite a few can fit in a tiny amount of semen.

15 During sexual arousal, the penis becomes hard and erect. What is happening to cause that? **c. Blood vessels are dilating to allow free flow of blood into the penis, while vascular shunts constrict drainage and cause blood pressure to rise in the corpora cavernosa.**

16 What happens during meiosis in the male? **b. The first division divides a diploid cell into two haploid cells, which then split again to form four sperm.** There's no such thing as a diploid sperm because as a sex cell, the sperm carries only half the regular complement of 46 chromosomes. And because another division takes place after the initial division in meiosis, the final product of the process is four cells, not two.

17 A man has 46 chromosomes in a spermatocyte. How many chromosomes are in each sperm? **b. 23.** Because the spermatocyte is the original cell that undergoes division, and because a human has a total of 46 chromosomes only *after* sperm meets egg, you must divide the number 46 in half.

18 Synapsis, or side-by-side pairing, of homologous chromosomes **c. occurs in meiosis.** Specifically, synapsis occurs during prophase I, which is the second stage of meiosis.

19 Anaphase I of meiosis is characterized by which of the following? **d. Homologous chromosomes separate and move poleward with centromeres intact.**

20 During oogenesis, the three nonfunctional cells produced are called **c. polar bodies.** They eventually disintegrate.

21 What happens during prophase II, or the second phase of meiosis? **d. Structures disappear from the nuclei, and spindles and asters appear in the nuclear region.** Remember that as the *second* phase of meiosis, two cells that emerged from the first phase must divide again and ultimately become four cells. The homologous chromatids in each of the cells don't need to duplicate again, however, because only 23 chromosomes end up in each of the final four cells. Instead, as the second meiotic division begins, they merely separate. That's what is happening when the spindles and asters appear.

22 Following is a summary of what should appear in your drawings and descriptions of the stages of meiosis. For further reference, check out Figure 13-3.

In the drawing for late prophase I, at least two pairs of homologous chromosomes should be grouped into tetrads (In truth, there are 23 pairs, but simplified illustrations tend to show just two). The description for prophase I should include reference to the tetrad formation. The drawing for metaphase I should show the equatorial plane (a center horizontal line) with the tetrads aligned along it. The illustration also should show spindles radiating from each pole, with the tetrads attached to them by their centromeres. The description should include reference to the equatorial plane, the poles, and the spindles.

The drawing for anaphase I should show the tetrads moving to the top and bottom of the cell along the spindles and the cytoplasm slowly beginning to divide. In telophase I, the division becomes more pronounced and two new nuclei form. As the process enters interkinesis, the cytoplasm pinches off into two cells.

During prophase II, which also is the start of the second meiotic division, the contracted and divided chromatids migrate toward a new equatorial plane. The drawing of metaphase II should show all chromatids aligned on the equatorial plane. For anaphase II, you should show the chromatids pulling apart into chromosomes and moving toward the poles. In the final stage, telophase II, you should draw new nuclei forming around the chromosomes.

Chapter 14

Carrying Life: The Female Reproductive System

Men may have quite a few hard-working parts in their reproductive systems (see Chapter 13), but women are the ones truly responsible for survival of the species (biologically speaking, anyway). The female body prepares for reproduction every month for most of a woman's adult life, producing a secondary oocyte and then measuring out delicate levels of hormones to prepare for nurturing a developing embryo. When a fertilized ovum fails to show up, the body hits the biological reset button and sloughs off the uterine lining before building it up all over again for next month's reproductive roulette. But that's nothing compared to what the female body does when a fertilized egg actually settles in for a nine-month stay. Strap yourselves in for a tour of the incredible female baby-making machinery — practice questions included.

Identifying the Female Reproductive Parts and Their Functions

First and foremost in the female reproductive repertoire are the two *ovaries,* which usually take a turn every other month to produce a single ovum (secondary oocyte). Roughly the size and shape of large, unshelled almonds, the female gonads lie on either side of the uterus, below and slightly behind the *Fallopian tubes* (also called the *uterine tubes*). Each ovary has a *stroma* (body) of connective tissue surrounded by a dense fibrous connective tissue called the *tunica albuginea* (literally "white covering"); yes, that's the same name as the tissue surrounding the testes. In fact, the ovaries in a female and the testes in a male are *homologous,* meaning that they share similar origins.

External to the tunica albuginea is a layer of cuboidal cells known as the *germinal epithelium.* Over time, a mass of epithelial cells surrounds the *primary oocyte,* forming *primordial follicles,* or *primary follicles,* that become separated from the main body of the ovary. The ovaries of a young girl contain approximately 700,000 of these follicles, most of them present at birth. During growth of the ovary in a female fetus prior to birth, the fetal gonads (ovaries) contain primordial germ cells that produce an *oogonium* (diploid cell) that becomes the primary oocyte (diploid). This primary oocyte undergoes the first meiotic division, producing two haploid cells. One is the secondary oocyte (haploid). The other cell is a nonfunctional polar body. The secondary oocyte is arrested at metaphase II of meiosis until fertilization occurs. At fertilization, the secondary oocyte completes the remaining stages of meiosis, resulting in the formation of an ootid that develops into the mature female gamete, or *ovum*.

At puberty and approximately once each month until menopause (which we cover in the later section "Growing, Changing, and Aging"), the following happens:

1. **The *pars distal* (anterior lobe) of the *hypophysis* (pituitary gland) secretes *follicle-stimulating hormones*, or *FSH,* which prompt about 1,000 of the primordial follicles to resume cellular division by mitosis.**

   Usually, only one follicle matures to become a *Graafian follicle*. Non-identical, or *fraternal,* twins, triplets, or even more fetuses result if more than one follicle matures to the point of releasing an ovum (secondary oocyte).

2. **One cell of this mass, the *primary oocyte* (produced by oogenesis, or meiosis), becomes the secondary oocyte while the remaining cells surround it as part of the *cumulus oophorus* and others line the fluid-filled follicular cavity as the *membrana granulosa*.**

3. **As the ovum matures, its follicle moves toward the ovary's surface and begins secreting the hormone *estrogen,* which signals the *endometrium* (uterine lining) to thicken in preparation for implantation of a fertilized ovum, or zygote.**

4. **As blood levels of estrogen begin to rise, the pituitary stops releasing FSH and begins releasing *luteinizing hormone,* or *LH,* which prompts the Graafian follicle now at the surface of the ovary to rupture, triggering *ovulation* — the release of the secondary oocyte, more commonly referred to as an egg cell.**

5. **Ringed by follicular cells in what's called the *corona radiata,* the secondary oocyte enters the *coelom* (body cavity) and is swept into the *Fallopian tube* by fingerlike projections called *fimbriae* (Latin meaning "fringe").**

6. **It takes approximately three days for the ovum to travel through the Fallopian tube to the uterus.**

Meanwhile, back at the ovary, a clot has formed inside the ruptured follicle and the membrana granulosa cells are replaced by yellow luteal cells, forming a *corpus luteum* (literally "yellow body") on the surface of the ovary. This new endocrine gland secretes *progesterone,* a hormone that signals the uterine lining to prepare for possible implantation of a fertilized egg, inhibits the maturing of Graafian follicles, ovulation, and the production of estrogen to prevent menstruation; and stimulates further growth in the mammary glands (which is why some women get sore breasts a few days before their periods begin). If pregnancy occurs, the placenta also will release progesterone to prevent menstruation throughout the pregnancy.

If the ovum (secondary oocyte) isn't fertilized, the corpus luteum dissolves after 10 to 14 days to be replaced by scar tissue called the *corpus albicans*. If pregnancy does occur, the corpus luteum remains and grows for about six months before disintegrating. Only about 400 of a woman's primordial follicles ever develop into secondary oocytes for the trip to the uterus. The rest ripen to various stages before degenerating into what are known as *atretic follicles* (or *corpora atretica*) over the course of her lifetime. Figure 14-1 shows what happens inside an ovary.

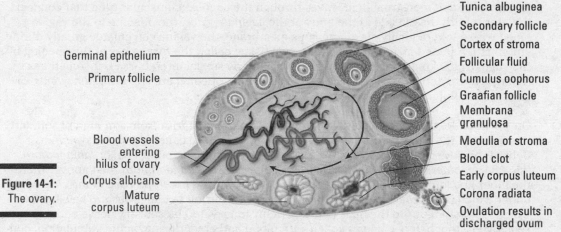

Tunica albuginea
Secondary follicle
Cortex of stroma
Follicular fluid
Cumulus oophorus
Graafian follicle
Membrana granulosa
Medulla of stroma
Blood clot
Early corpus luteum
Corona radiata
Ovulation results in discharged ovum

Germinal epithelium
Primary follicle
Blood vessels entering hilus of ovary
Corpus albicans
Mature corpus luteum

Figure 14-1:
The ovary.

Illustration by Imagineering Media Services Inc.

Fallopian tubes, oviducts, uterine tubes — call them what you will, but they are where the real business of fertilization takes place. Why? Because an egg must be fertilized within 24 hours of its release from the ovary to remain viable. These small, muscular tubes lined with *cilia* are nearly 5 inches long and, somewhat surprisingly, aren't directly connected to the ovaries. Instead, the funnel-shaped end, the *infundibulum,* of a tube is fragmented into ciliated fingerlike projections called *fimbriae* that help to move the secondary oocyte from the body cavity into the tube. When it's in the tube, where fertilization takes place, the combined motions of both cilia and peristalsis (the same muscle contractions that move food through the digestive system) propel the fertilized egg toward the uterus for implantation. If a fertilized egg implants anywhere else — say in the lining of the Fallopian tube itself — the pregnancy is referred to as *ectopic* and the woman must have immediate surgery to remove the developing embryo before it can damage the Fallopian tube and cause other internal injuries.

Although not attached to the ovaries, the Fallopian tubes are attached to the pear shaped *uterus,* which is located between the urinary bladder and the rectum. The upper, wide end of the uterus is called the *fundus;* the lower, narrow end that opens into the vagina is the *cervix;* and the central region is the *body.* The upper end of the vaginal canal surrounds the cervix of the uterus, forming a recess call the *vaginal fornix.* The largest portion of the recess lies dorsal to the cervix and is called the *posterior fornix.* The endometrium (mucous membrane) lines the uterus in varying amounts depending on the stage of a woman's menstrual cycle or pregnancy. The endometrium is the functional layer undergoing changes during the menstrual cycle. Each month, it thickens and becomes engorged with blood in preparation for receiving a fertilized egg and the onset of gestation. If the secondary oocyte isn't fertilized, the functional layer of the endometrium is sloughed off as menstrual flow. This lining is supported by a thick muscular layer called the *myometrium,* which is under the control of the autonomic nervous system and comes into play when the uterus contracts, such as during labor.

There are three phases in the uterine cycle:

✔ **Menstrual:** Shedding of the endometrium

✔ **Proliferation:** Rebuilding of the endometrium

✔ **Secretory:** Enriching the blood supply after ovulation to provide the nutrients that prepare the endometrium for the blastocyst.

Sperm enter and menstrual fluid leaves through the *vagina,* a muscular tube that connects the uterus with the outside of the body at the *vaginal orifice,* the opening to the vagina. Lined with a fold of highly elastic mucous membrane, the vagina can enlarge greatly during childbirth. A folded membrane of connective tissue called the *hymen* lies at the opening of the vaginal canal until it is ruptured or torn, often by sexual intercourse but sometimes by other physical activities. At either side of the vaginal opening are two *Bartholin glands,* or greater vestibular glands, that secrete a lubricating mucus.

On the outside, the female genitalia extend toward the posterior from a mound of soft, fatty tissue called the *mons pubis* that covers the bone structure called the *pubic symphysis.* Behind this, the *vulva* consists of two flaps of fatty tissue: the outer lips, or *labia majora* (singular: *labium*); and the smaller, hairless inner lips, the *labia minora.* Just above where the inner lips join is a small flap of tissue called the *clitoral hood (prepuce of the clitoris),* under which is the *clitoris,* erectile tissue that swells during sexual arousal. The clitoris, which is homologous to the male penis, is richly innervated by sensory nerve endings and sensitive to touch. It becomes swollen with blood and erect during tactile stimulation, contributing to the female's sexual arousal. Below the clitoris is the external opening of the urethra and below that is the *introitus,* or vaginal orifice. Posterior to the vagina is the anus, the opening for the rectum.

Find out how familiar you are with the female anatomy:

1. What role does FSH play in the female reproductive system?

   a. Released in the wake of ovulation, it signals the uterus to prepare for a fertilized egg.

   b. Each of these fingerlike projections working together sweeps the egg into the Fallopian tube.

   c. Secreted by the pituitary gland, it prompts the resumption of meiosis in the ovaries.

   d. It prompts the Graafian follicle to rupture.

2. What does it mean to say that the Graafian follicle is maturing?

   a. One of about 1,000 primordial follicles has developed enough to release a secondary oocyte.

   b. The secondary oocyte has been released and scar tissue is forming around the remaining follicles.

   c. The corona radiata is interacting with the fimbriae.

   d. The fertilized ovum is preparing to implant in the endometrium.

3. Which one of the following is not a function of estrogen?

   a. Preparing the endometrium

   b. Supporting development of the secondary oocyte

   c. Supporting development of the corpus luteum

   d. Preventing secretion of FSH from the pituitary

4. Which of the following does not produce a hormone?

   a. Corpus luteum

   b. Pars distalis

   c. Graafian follicle

   d. Corpus albicans

5.–9. Match the term to its description.

   5. _____ Ectopic   **a.** Period of intrauterine development

   6. _____ Gestation   **b.** Cessation of menses

   7. _____ Ovulation   **c.** Development of out-of-place embryo

   8. _____ Menopause   **d.** Release of secondary oocyte into coelom

   9. _____ Luteinization   **e.** Glandular development by membrana granulosa

10. Why does the corpus luteum produce progesterone?

   a. To stimulate development of the cumulus oophorus

   b. To trigger ovulation

   c. To prepare a woman's system for pregnancy and prevent menstruation

   d. To prepare the infundibulum to fragment into fimbriae

11.–15. Match the term to its description.

   11. _____ Corona radiata   **a.** Fingerlike projections at the end of a Fallopian tube

   12. _____ Endometrium   **b.** Lining of the follicle

   13. _____ Fimbriae   **c.** Granulosa cells surrounding the secondary oocyte

   14. _____ Stroma   **d.** Body of the ovary

   15. _____ Membrana granulosa **e.** Inner lining of the uterus

16.–19. Match the description to the hormone (one answer is used twice).

   16. _____ Prevents menstruation in pregnant females   **a.** Progesterone

   17. _____ Triggers ovulation   **b.** Luteinizing hormone (LH)

   18. _____ Secretion of the developing follicle   **c.** Estrogen

   19. _____ Secreted by the corpus luteum

20.–35. Use the terms that follow to identify the anatomy of the female reproductive system shown in Figure 14-2.

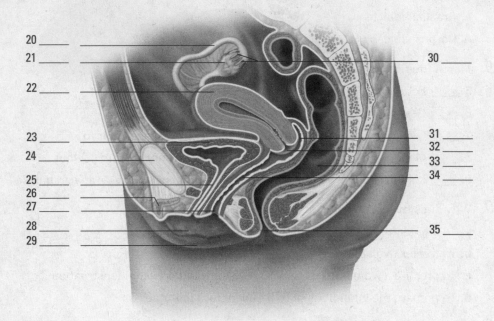

20 _____ _____
21 _____ _____
22 _____ _____
23 _____ _____
24 _____ _____
25 _____ _____
26 _____ _____
27 _____ _____
28 _____ _____
29 _____ _____
30 _____
31 _____
32 _____
33 _____
34 _____
35 _____

Figure 14-2:
The female reproductive system.

Illustration by Imagineering Media Services Inc.

a. Cervix

b. Fimbriae

c. Urinary bladder

d. Clitoris

e. Labium major

f. Ovary

g. Rectum

h. Posterior fornix

i. Vaginal orifice

j. Anus

k. Symphysis pubis

l. Uterus

m. Labium minor

n. Uterine (Fallopian) tube

o. Vagina

p. Urethra

Making Eggs: A Mite More Meiosis

We cover meiosis in detail in Chapter 13, but we also cover the process in this section, exploring how meiosis contributes to making a *zygote* (fertilized egg) and its chromosomal makeup. Normal human diploid, or 2N, cells contain 23 pairs of homologous chromosomes for a total of 46 chromosomes each. One chromosome from each pair comes from the individual's mother, and the other comes from the father. Each homologous pair contains the same type of genetic information, but the *expression* of this genetic information may differ from one chromosome of the pair to the homologous chromosome. For example, one chromosome from the mother may carry the genetic coding for blue eyes, whereas the homologous chromosome from the father may code for brown eyes.

Although all the body's cells have 46 chromosomes — including the cells that eventually mature into the ovum and the sperm — each gamete (either ovum or sperm) must offer only half that number if fertilization is to succeed. That's where meiosis steps in. As you see in Figure 14-3, the number of chromosomes is cut in half during the first meiotic division, producing gametes that are haploid, or 1N. The second meiotic division produces four haploid sperms in the male but only one functional haploid secondary oocyte in the female. When fertilization occurs, the new zygote will contain 23 pairs of homologous chromosomes for a total of 46 chromosomes. The zygote then proceeds through mitosis to produce the body's cells, distributing copies of all 46 chromosomes to each new cell.

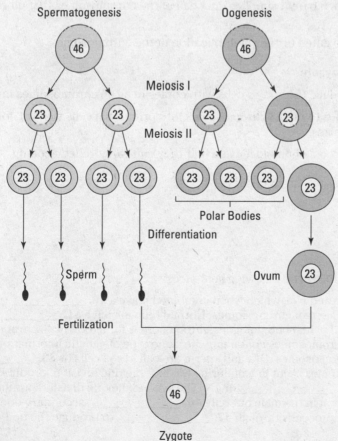

Figure 14-3:
How chromosomes divide up in meiosis.

Think you've conquered this process? Find out by answering the following practice questions:

EXAMPLE

Q. How many chromosomes are in most of the cells in your body?

   a. 46

   b. 23

   c. 92

A. The correct answer is 46. That's the usual human complement.

36. Another term for a sperm or an ovum is

   a. a zygote.

   b. a blastocyst.

   c. a gamete.

37. Why are there only half as many chromosomes in each ovum or sperm as there are in the cells of the rest of the body?

   a. The cell's nucleus isn't large enough to hold more.

   b. During fertilization each parent supplies only half of the chromosomes that ultimately will develop into a child.

   c. Angiogenesis has prevented further multiplication of the chromosomes.

38. What happens inside a zygote?

   a. Mitosis begins to produce the body's cells with copies of all 46 chromosomes in place.

   b. Gene expression begins to put 23 chromosomes into primordial cells that will join together to make 46 chromosomes.

   c. The chromosomes become homologous so that they can carry genetic coding.

39. How many chromosomes are in a haploid cell?

   a. 46

   b. 23

   c. 92

40.–47. Fill in the blanks to complete the following sentences:

Meiosis produces sperm and ova, which when combined make a **40.** _____ (fertilized egg) with its full complement of chromosomes. Normal humans have **41.** _____, or 2N, cells containing 23 pairs of homologous chromosomes for a total of 46 chromosomes each. A pair of chromosomes containing the same type of genetic information are **42.** _____ chromosomes. Ova and sperm are called sex cells or **43.** _____. The number of chromosomes is cut in half during the first meiotic division, producing gametes that are **44.** _____, or 1N. The second meiotic division produces **45.** _____ sperm in the male but only **46.** _____ secondary oocyte in the female. The zygote then proceeds through **47.** _____ to produce the body's cells.

Making Babies: An Introduction to Embryology

Fertilization (the penetration of a secondary oocyte by a sperm) occurs in the Fallopian tube. Fertilization must occur within 24 hours of ovulation or the secondary oocyte degenerates. But that doesn't mean that sexual intercourse outside that time frame can't lead to pregnancy. Research indicates that spermatozoa can survive up to seven days inside the uterus and Fallopian tubes. If a sperm is still motile — that is, if it's still whipping its flagellum tail — when an ovum comes down the tube, it will do what it was made to do and penetrate the ovum's membrane.

The sperm releases enzymes that allow it to digest its way into the secondary oocyte, leaving its flagellum behind. After that first sperm penetration, the membrane surrounding the ovum instantly solidifies, preventing any other sperm from getting inside. With the penetration of the sperm, meiosis II is completed, resulting in an ovum and a second polar body. When the pronuclei of the ovum and the sperm unite, presto! You've got a zygote (fertilized egg). Over the next three to five days, the zygote moves through the Fallopian tube to the uterus, undergoing *cleavage* (mitotic cell division) along the way: Two cells become four smaller cells, four cells become eight smaller cells, and then those eight cells become a solid 16-cell ball called a *morula*. After five days of cleavage, the cells form a hollow ball called a *blastula,* or *blastocyst.* The inner hollow region is called the *blastocoele,* and the outer-layer cells are called the *trophoblast.* Figure 14-4 illustrates this process of development.

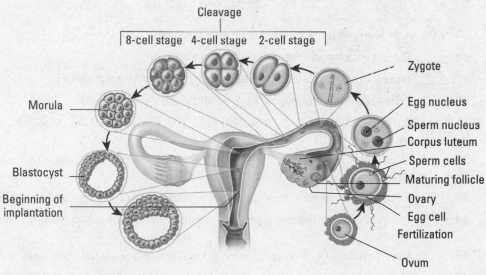

Figure 14-4:
Embryonic development.

Illustration by Imagineering Media Services Inc.

Within three days of its arrival in the uterine cavity (generally within a week of fertilization), the blastocyst implants in the endometrium, and some of the blastocyst's cells — called *totipotent embryonic stem cells* — organize into an inner cell mass also known as the *embryonic disk,* or *embryoblast. Totipotent* describes the ability of a cell to divide and produce any of the body's differentiated cells. Over time, the embryonic disk *differentiates* into the tissues of the developing embryo. Cells above the disk form the *amnion* and its resulting *amniotic cavity,* and those below form the *gut cavity* and two primitive *germ layers.* The layer nearest the amniotic cavity forms the *ectoderm* while that nearest the gut cavity forms the *endoderm.* Between the two layers, additional ectodermal cells develop to form a third layer, the *mesoderm. Pluripotent stem cells* are derived from the inner cell mass (also known as the *embryoblast* or *pluriblast*) found in the *blastocyst cavity* and can produce the different tissues of the body. The ectoderm forms skin and nerve tissue; the mesoderm forms bones, cartilage, connective tissue, muscles, and organs; and the endoderm forms the linings of the organs and glands.

To keep these terms straight, remember that *endo*– means "inside or within," *ecto*– means "outer or external," and *meso*– means "middle."

After three weeks of development, the heart begins beating. In the fourth week of development, the embryonic disk forms an elongated structure that attaches to the developing placenta by a connecting stalk. A head and jaws form while primitive buds sprout; the buds will develop into arms and legs. During the fifth through seventh weeks, the head grows rapidly and a face begins to form (eyes, nose, and a mouth). Fingers and toes grow at the ends of the elongating limb buds. All internal organs have started to form. After eight weeks of development, the embryo begins to have a more human appearance and is referred to as a *fetus.*

The outer cells of the blastocyst, together with the endometrium of the uterus, form the *placenta,* a new internal organ that exists only during pregnancy. The placenta attaches the fetus to the uterine wall, exchanges gases and waste between the maternal and fetal bloodstreams, and secretes hormones to sustain the pregnancy.

Now that you've refreshed your memory a bit about how babies are made (beyond the birds and bees talks), try the following practice questions:

48.–57. Mark the statement with a T if it's true or an F if it's false:

48. _____ The mesoderm will form nerve tissue and skin.

49. _____ The embryonic stage is complete at the end of the eighth week.

50. _____ Cells above the embryonic disk form the gut cavity.

51. _____ During the fifth through seventh weeks, the arm and leg buds elongate, and fingers and toes begin to form.

52. _____ Cleavage is successive mitotic divisions of the embryonic cells into smaller and smaller cells.

53. _____ As the zygote moves through the Fallopian tube, it undergoes meiosis.

54. _____ The placenta exchanges gases and waste between the maternal blood and the fetal blood.

55. _____ After five days of cleavage, the cells form into a hollow ball called the morula.

56. _____ The embryonic stage is complete at the end of the fifth week of development.

57. _____ Sexual intercourse five days before ovulation cannot lead to pregnancy.

Growing from Fetus to Baby

Pregnancy is divided into three periods called *trimesters* (although many new parents bemoan a postnatal fourth trimester until the baby sleeps through the night). The first 12 weeks of development mark the first trimester, during which *organogenesis* (organ formation) is established. During the second trimester (typically considered to be weeks 13 to 27), all fetal systems continue to develop and rapid growth triples the fetus's length. By the third trimester (typically considered to be weeks 28 to 40), all organ systems are functional and the fetus usually is considered *viable* (capable of surviving outside the womb) even if it's born prematurely. The overall growth rate slows in the third trimester, but the fetus gains weight rapidly.

Milestones in fetal development can be marked monthly.

- At the **end of the second month** (when the terminology changes from "embryo" to "fetus"), the head remains overly large compared to the developing body, and the limbs are still short. All major regions of the brain have formed.

- During the **third month,** body growth accelerates and head growth slows. The arms reach the length they will maintain during fetal development. The bones begin to ossify, and all body systems have begun to form. The circulatory (cardiovascular) system supplies blood to all the developing extremities, and even the lungs begin to practice "breathing" amniotic fluid. By the end of the third month, the external genitalia are visible in the male (ultrasound technicians call this a "turtle sign"). The fetus is a bit less than 4 inches long and weighs about 1 ounce.

- The body grows rapidly during the **fourth month** as legs lengthen, and the skeleton continues to ossify as joints begin to form. The face looks more human. The fetus is about 7 inches long and weighs 4 ounces.

- Growth slows during the **fifth month,** and the legs reach their final fetal proportions. Skeletal muscles become so active that the mother can feel fetal movement. Hair grows on the head, and *lanugo,* a profusion of fine soft hair, covers the skin. The fetus is about 12 inches long and weighs ½ to 1 pound.

- The fetus gains weight during the **sixth month,** and eyebrows and eyelashes form. The skin is wrinkled, translucent, and reddish because of dermal blood vessels. The fetus is between 11 and 14 inches long and weighs a bit less than 1½ pounds.

- During the **seventh month,** skin becomes smoother as the fetus gains subcutaneous fat tissue. The eyelids, which are fused during the sixth month, open. Usually, the fetus turns to an upside-down position. It's between 13 and 17 inches long and weighs from 2½ to 3 pounds.

- During the **eighth month,** subcutaneous fat increases, and the fetus shows more baby-like proportions. The testes of a male fetus descend into the scrotum. The fetus is now 16 to 18 inches long and has grown to just under 5 pounds.

- During the **ninth month,** the fetus plumps up considerably with additional subcutaneous fat. Much of the lanugo is shed, and fingernails extend all the way to the tips of the fingers. The average newborn at the end of the ninth month is 20 inches long and weighs about 7½ pounds.

The following practice questions deal with the development of the fetus during its 40 weeks in the womb:

58. What needs to happen before a fetus can be considered viable?

a. The heart rate must be easily detected.

b. All organ systems must be functional.

c. Subcutaneous fat must be built up to at least an inch.

d. The fetus must have turned into an upside-down position.

59. Describe one new fetal development for each month:

3rd month: _____

4th month: _____

5th month: _____

6th month: _____

7th month: _____

8th month: _____

9th month: _____

Growing, Changing, and Aging

After a baby arrives, the female reproductive system goes into a different form of intense activity. Throughout the pregnancy, the placenta has been producing estrogen and progesterone to sustain the fetus. But after the baby is born, the sudden drop in hormonal blood levels triggers the pituitary gland to release *prolactin,* a hormone that stimulates the woman's mammary glands to produce milk. An additional hormone from the pituitary gland, *oxytocin,* stimulates the release of milk from the breast in a process called *lactation.* First, however, the glands produce *colostrum,* a thin, yellowish fluid rich in antibodies and minerals to sustain a newborn. Both colostrum and milk flow from a number of lobes inside the breast through *lactiferous ducts* that converge on the nipple.

From birth to 4 weeks of age, the newborn is called a *neonate.* Faced with survival after its physical separation from the mother, a neonate must abruptly begin to process food, excrete waste, obtain oxygen, and make circulatory adjustments.

From 4 weeks to 2 years of age, the baby is called an *infant.* Growth during this period is explosive under the stimulation of circulatory growth hormones from the pituitary gland, adrenal steroids, and thyroid hormones. The infant's *deciduous* teeth, also called *baby* or *milk teeth,* begin to form and erupt through the gums. The nervous system advances, making coordinated activities possible. The baby begins to develop language skills.

From 2 years to puberty, you're looking at a *child.* Influenced by growth hormones, growth continues its rapid pace as deciduous teeth are replaced by permanent teeth. Muscle coordination, language skills, and intellectual skills also develop rapidly.

From puberty, which starts between the ages of 11 and 14, to adulthood, the child is called an *adolescent.* Growth occurs in spurts. Girls achieve their maximum growth rate between the ages of 10 and 13, whereas boys experience their fastest growth between the ages of 12 and 15. Primary and secondary sex characteristics begin to appear. Growth terminates when the epiphyseal plates of the long bones ossify sometime between the ages of 18 and 21 (see Chapter 5 for an introduction to the bones of the body). Motor skills and intellectual abilities continue to develop, and psychological changes occur as adulthood approaches.

The young adult stage covers roughly 20 years, from the age of 20 to about 40. Physical development reaches its peak and adult responsibilities are assumed, often including a career, marriage, and a family. After about age 30, physical changes that indicate the onset of aging begin to occur.

From age 40 to about 65, physiological aging continues. Gray hair, diminished physical abilities, and skin wrinkles are outward signs of aging. Women go through *menopause,* the cessation of monthly cycles, which is also known as *climacteric* or the *change of life.* While *menarche* (the onset of menstruation) may begin any time between 10 and 15 years of age, the female body's monthly reproductive cycle slows and stops entirely between the ages of 40 and 55. With the cessation of menses comes a decrease in size of the uterus, shortening of the vagina, shrinkage of the mammary glands, disappearance of Graafian follicles, and shrinkage of the ovaries. For about six years prior to menopause, many women experience a stage called *perimenopause* during which increasingly irregular hormone secretions can cause fluctuations in menstruation and a sensation called *hot flashes.*

From age 65 until death is the period of *senescence,* the process of aging. Individual adults can show widely varying patterns of aging in part because of differences in genetic background and physical activities. Signs of senescence include

- Loss of skin elasticity and accompanying sagging or wrinkling
- Weakened bones and decreasingly mobile joints
- Weakened muscles
- Impaired coordination, memory, or intellectual function
- Cardiovascular problems
- Reduced immune responses
- Decreased respiratory function caused by reduced lung elasticity
- Decreased peristalsis and muscle tone in the digestive and urinary tracts

Following are some practice questions on the aging cycle:

60. When the pituitary gland releases prolactin, which of the following glands responds?

a. Parathyroid gland

b. Mammary gland

c. Pars distalis

d. Adrenal gland

61. An individual's height stops increasing

a. during prepubescence.

b. upon onset of menopause.

c. when the epiphyseal plates of the long bones ossify.

d. during middle age.

62. What is the formal term for a 20-month-old child?

a. Baby

b. Infant

c. Toddler

d. Neonate

63.–69. Match the description to its life stage.

63. _____ Assumes adult responsibilities, possibly including marriage and a family

64. _____ Faced with survival, it must process food, excrete waste, and obtain oxygen

65. _____ Primary and secondary sex characteristics begin to appear

66. _____ Experiences senescence

67. _____ Deciduous teeth begin to form

68. _____ Women go through menopause

69. _____ From 2 years of age to puberty

a. Neonate

b. Infant

c. Child

d. Adolescent

e. Young adult

f. Middle-aged adult

g. Old adult

Answers to Questions on the Female Reproductive System

The following are answers to the practice questions presented in this chapter.

1. What role does FSH play in the female reproductive system? **c. Secreted by the pituitary gland, it prompts the resumption of meiosis in the ovaries.**

   FSH stands for follicle stimulating hormone, and the Graafian follicle that ultimately releases the egg results from meiosis within the ovary.

2. What does it mean to say that the Graafian follicle is maturing? **a. One of about 1,000 primordial follicles has developed enough to release a secondary oocyte.** In biology, "primordial" is defined as a cell or tissue in the earliest stage of development, so that immature follicle must mature into a Graafian follicle before it can rupture and release the egg cell.

3. Which one of the following is not a function of estrogen? **c. Supporting development of the corpus luteum.** By the time the corpus luteum starts to develop, estrogen already has begun to bow out of the reproductive equation.

4. Which of the following does not produce a hormone? **d. Corpus albicans.** Corpus albicans is the scar tissue that follows disintegration of the corpus luteum, so it makes sense that it doesn't produce any hormones.

5. Ectopic: **c. Development of out-of-place embryo.** *Ektopos* in Greek literally means "out of place."

6. Gestation: **a. Period of intrauterine development.** *Gestation* is just a fancy term for pregnancy, and intra– means "within."

7. Ovulation: **d. Release of secondary oocyte into coelom.** Ignore the reference to the body cavity. "Release of secondary oocyte" is the giveaway here.

8. Menopause: **b. Cessation of menses.** Menses is the same as menstruation.

9. Luteinization: **e. Glandular development by membrana granulosa.** The granulosa cells become luteal cells to form the corpus luteum, which is actually a new gland each month.

10. Why does the corpus luteum produce progesterone? **c. To prepare a woman's system for pregnancy and prevent menstruation.** It stands to reason that if the corpus luteum forms after the Graafian follicle ruptures, the progesterone it's producing is signaling the uterus that a fertilized egg may be on its way.

11. Corona radiata: **c. Granulosa cells surrounding the secondary oocyte**

12. Endometrium: **e. Inner lining of the uterus**

13. Fimbriae: **a. Fingerlike projections at the end of a Fallopian tube**

14. Stróma: **d. Body of the ovary**

15. Membrana granulosa: **b. Lining of the follicle**

16. Prevents menstruation in pregnant females: **a. Progesterone**

17. Triggers ovulation: **b. Luteinizing hormone (LH)**

18. Secretion of the developing follicle: **c. Estrogen**

19. Secreted by the corpus luteum: **a. Progesterone**

20–35 Following is how Figure 14-2, the female reproductive system, should be labeled.

20. **n. Uterine (Fallopian) tube;** 21. **f. Ovary;** 22. **l. Uterus;** 23. **c. Urinary bladder;** 24. **k. Symphysis pubis;** 25. **p. Urethra;** 26. **d. Clitoris;** 27. **i. Vaginal orifice;** 28. **m. Labium minor;** 29. **e. Labium major;** 30. **b. Fimbrae;** 31. **h. Posterior fornix;** 32. **a. Cervix;** 33. **g. Rectum;** 34. **o. Vagina;** 35. **j. Anus**

36 Another term for a sperm or an ovum is **c. a gamete.** You already know that a zygote and a blastocyst are embryonic forms, so gamete is the only possible correct answer.

37 Why are there only half as many chromosomes in each ovum or sperm as there are in the cells of the rest of the body? **b. During fertilization each parent supplies only half of the chromosomes that ultimately will develop into a child.** Again, if you have any doubt, you can reach the correct answer through a process of elimination. To say that the cell's nucleus isn't large enough to hold more is ridiculous. And angiogenesis has nothing to do with any of these processes. Therefore, the only logical answer must be the correct one.

38 What happens inside a zygote? **a. Mitosis begins to produce the body's cells with copies of all 46 chromosomes in place.** A zygote will develop into an embryo, and every cell of the developing embryo needs a full complement of chromosomes.

39 How many chromosomes are in a haploid cell? **b. 23.**

TIP

The number of chromosomes in a haploid cell may be easier to remember this way: *Haploid* is *half,* but *diploid* is *double* that.

40–47 Meiosis produces sperm and ova, which when combined make a **40. zygote** (fertilized egg) with its full complement of chromosomes. Normal humans have **41. diploid,** or 2N, cells containing 23 pairs of homologous chromosomes for a total of 46 chromosomes each. A pair of chromosomes containing the same type of genetic information are **42. homologous** chromosomes. Ova and sperm are called sex cells or **43. gametes.** The number of chromosomes is cut in half during the first meiotic division, producing gametes that are **44. haploid,** or 1N. The second meiotic division produces **45. four haploid** sperm in the male but only **46. one functional haploid** secondary oocyte in the female. The zygote then proceeds through **47. mitosis** to produce the body's cells.

48 The mesoderm will form nerve tissue and skin. **False.**

49 The embryonic stage is complete at the end of the eighth week. **True.**

50 Cells above the embryonic disk form the gut cavity. **False.**

51 During the fifth through seventh weeks, the arm and leg buds elongate, and fingers and toes begin to form. **True.**

52 Cleavage is successive mitotic divisions of the embryonic cells into smaller and smaller cells. **True.**

53 As the zygote moves through the Fallopian tube, it undergoes meiosis. **False.**

54 The placenta exchanges gases and waste between the maternal blood and the fetal blood. **True.**

55 After five days of cleavage, the cells form into a hollow ball called the morula. **False.**

56 The embryonic stage is complete at the end of the fifth week of development. **False.**

57 Sexual intercourse five days before ovulation cannot lead to pregnancy. **False.** It certainly can happen — and does.

58 What needs to happen before a fetus can be considered viable? **b. All organ systems must be functional.** This generally occurs by the third trimester, but even then a premature birth can have serious health consequences for the newborn.

59 Describe one new fetal development for each month:

3rd month: Bones begin to ossify, body growth accelerates while head growth slows, lungs begin to practice breathing amniotic fluid, external genitalia visible in male, 4 inches long and about 1 ounce

4th month: Body grows rapidly, legs lengthen, joints begin to form, face looks more human, roughly 7 inches long and 4 ounces

5th month: Mother can feel fetal movement, hair grows on head, lanugo covers skin, about 12 inches long and ½ to 1 pound

6th month: Eyebrows and eyelashes form, weight gain accelerates, roughly 11 to 14 inches long and just under 1½ pounds

7th month: Subcutaneous fat begins to form, eyelids open, usually turns to upside-down position, between 13 and 17 inches long and 2½ to 3 pounds

8th month: Subcutaneous fat increases, fetus appears more "babylike," testes of male descend into scrotum, 16 to 18 inches long and roughly 5 pounds

9th month: Substantial "plumping" with subcutaneous fat, lanugo shed, fingernails extend to fingertips, average newborn is 20 inches long and weighs 7½ pounds

60 When the pituitary gland releases prolactin, which of the following glands responds? **b. Mammary gland.** Everybody say "moo"!

61 An individual's height stops increasing **c. when the epiphyscal plates of the long bones ossify.** Just so you know, that occurs sometime between the ages of 18 and 21.

62 What is the formal term for a 20-month-old child? **b. Infant.** The terms "baby" and "toddler" aren't part of the formal medical lexicon, and a neonate is less than 4 weeks old.

63 Assume adult responsibilities, possibly including marriage and a family: **e. Young adult.**

64 Faced with survival, it must process food, excrete waste, and obtain oxygen: **a. Neonate.**

65 Primary and secondary sex characteristics begin to appear: **d. Adolescent.**

66 Experiences senescence: **g. Old adult.**

67 Deciduous teeth begin to form: **b. Infant.**

68 Women go through menopause: **f. Middle-aged adult.**

69 From 2 years of age to puberty: **c. Child.**

Part V
Mission Control: All Systems Go

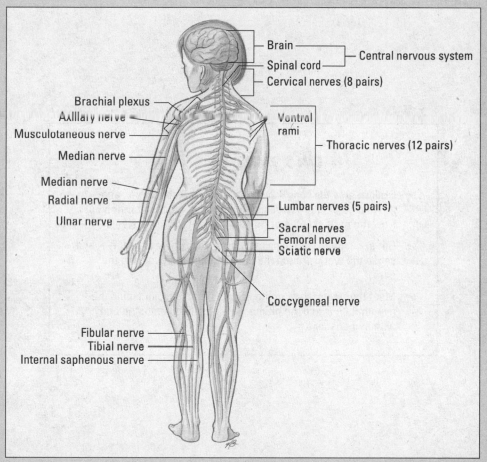

Illustration by Kathryn Born, MA

Check out the Cheat Sheet for some Latin and Greek roots, prefixes, and suffixes that will make your memorization of anatomical terms easier: www.dummies.com/extras/ anatomyphysiologywb.

In this part . . .

✔ Explore how the body's systems let it move, think, feel, and thrive on a day-to-day basis. Build up nerves; then branch out from the central to the peripheral and autonomic systems.

✔ Get acquainted with mission control — the human brain — and sense the world around the body through the eyes and the ears.

✔ Meet the ringmasters of the endocrine system, including the hypothalamus and the pituitary gland, and balance the body with homeostasis.

Chapter 15

Feeling Jumpy: The Nervous System

Throughout this book, you look at the human body from head to toe, exploring how it collects and distributes the molecules it needs to grow and thrive, how it reproduces itself, and even how it gets rid of life's nastier byproducts. In this chapter, however, you look at the living computer that choreographs the whole show, the one system that contributes the most to making us who we are as humans: the nervous system.

In this chapter, you get a feel for how the nervous system is put together. You practice identifying the parts and functions of nerves and the brain itself as well as the structure and activities of the Big Three parts of the whole nervous system: the central, the peripheral, and the autonomic systems. In addition, we touch on the sensory organs that bring information into the human body.

Building from Basics: Neurons, Nerves, Impulses, and Synapses

The nervous system is the communications network that goes into nearly every part of the body, innervating your muscles, pricking your pain sensors, and letting you reach beyond yourself into the larger world. More than 80 major nerves make up this intricate network, with billions of *neurons* (individual nerve cells) in the nervous system. It's through this complex network that you respond both to external and internal stimuli, demonstrating a characteristic called *irritability* (the capacity to respond to stimuli, not the tendency to yell at annoying people).

There are three functional types of cells in the nervous system: *receptor cells* that receive a stimulus (sensing); *conductor cells* that transmit impulses (integrating); and *effector cells,* or motor neurons, which bring about a response such as contracting a muscle. Put another way, there are three functions of the human nervous system as a whole:

✔ **Orientation:** The ability to generate nerve impulses in response to changes in the external and internal environments (this also can be referred to as *perception*)

✏ **Coordination:** The ability to receive, sort, and direct those signals to channels for response (this also can be referred to as *integration*)

✏ **Conceptual thought:** The capacity to record, store, and relate information received and to form plans for future reactions to environmental change (which includes specific *action*)

Before trying to study the system as a whole, it's best to break it down into building blocks first: neurons, nerves, impulses, and synapses.

Neurons

The basic unit that makes up nerve tissue is the *neuron* (also called a *nerve cell*). Its properties include that marvelous *irritability* that we speak of in the chapter introduction as well as *conductivity,* otherwise known as the ability to transmit a *nerve impulse.*

The central part of a neuron is the *cell body,* or *soma,* that contains a large nucleus with one or more nucleoli, mitochondria, Golgi apparatus, numerous ribosomes, and *Nissl bodies* that are associated with conduction of a nerve impulse. (See Chapter 2 for an overview of a cell's primary parts.) Extending from the cell body are threads of cytoplasm, or cytoplasmic projections, containing specialized fibrils, or *neurofibrillae.* Two types of these cytoplasmic projections play a role in neurons: *Dendrites* conduct impulses to the cell body while *axons* (nerve fibers) usually conduct impulses away from the cell body (see Figure 15-1). Each neuron has only one axon; however, each axon can have many branches called *axon collaterals,* enabling communication with many target cells. The point of attachment on the soma is called the *axon hillock.* In addition, each neuron may have one dendrite, several dendrites, or none at all.

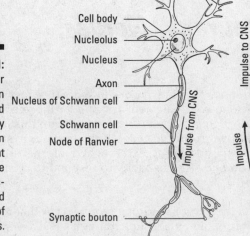

Figure 15-1: The motor neuron on the left and the sensory neuron on the right show the cell structures and the paths of impulses.

Illustration by Kathryn Born, MA

There are three types of neurons, as follows:

✔ **Motor neurons,** or *efferent neurons,* transmit messages from the brain and spinal cord to effector organs, including muscles and glands, triggering them to respond. Motor neurons are classified structurally as *multipolar* because they're star-shaped cells with a single, large axon and numerous dendrites.

✔ **Sensory neurons,** or *afferent neurons,* are triggered by physical stimuli, such as light, and pass the impulses on to the brain and spinal cord. Sensory fibers have special structures called receptors, or end organs, where the stimulus is propagated. Sensory neurons can be classified structurally as either *unipolar* or *bipolar.* Unipolar neurons have a single *process* (a projection or outgrowth of tissue) that divides shortly after leaving the cell body; one branch conveys impulses from sense organs while the other branch carries impulses to the central nervous system. Bipolar neurons have two processes — one dendrite and one axon.

✔ **Association neurons** (also called *internuncial neurons, interneurons,* or *intercalated neurons*) are triggered by sensory neurons and relay messages between neurons within the brain and spinal cord. Interneurons, like motor neurons, are classified structurally as multipolar.

Here are a couple of handy memory devices: Afferent connections arrive, and efferent connections exit. Dendrites deliver impulses while axons send them away.

Nerves

Whereas neurons are the basic unit of the nervous system, *nerves* are the cablelike bundles of axons that weave together the peripheral nervous system. There are three types of nerves:

✔ **Afferent nerves** are composed of sensory nerve fibers (axons) grouped together to carry impulses from receptors to the central nervous system.

✔ **Efferent nerves** are composed of motor nerve fibers carrying impulses from the central nervous system to effector organs, such as muscles or glands.

✔ **Mixed nerves** are composed of both afferent and efferent nerve fibers.

The diameter of individual axons (nerve fibers) tends to be microscopically small — many are no more than a micron, or one-millionth of a meter. But these same axons extend to lengths of 1 millimeter and up. The longest axons in the human body run from the base of the spine to the big toe of each foot, meaning that these single-cell fibers may be 1 meter or more in length.

Each axon is swathed in *myelin,* a white, fatty material made up of concentric layers of *Schwann cells* acting as electrical insulators in peripheral nerves. *Oligodendrocytes* (meaning cells with few branches) in the central nervous system are also associated with myelinated nerve fibers. The result is a structure referred to as a *myelin sheath.* Gaps in the sheath called *nodes of Ranvier* give the underlying nerve fiber access to extracellular fluid, to speed up propagation of the nerve impulse. *Non-myelinated nerve fibers* lie within body organs and therefore don't need protective myelin sheaths to help them transmit impulses. Many peripheral nerve cell fibers also are protected by a *neurilemmal sheath,* a membrane that surrounds both the nerve fiber and its myelin sheath.

From the inside out, nerves are composed of the following:

- **Axon:** The impulse-conducting process of a neuron

- **Myelin sheath:** A dielectric (polarizable) insulating envelope that protects the nerve fiber and facilitates transmission of nerve impulses by reducing charge leakage and loss of impulse through increasing resistance and reducing capacitance (the charge storage coefficient) by the way in which the layers are connected

- **Neurolemma (or neurilemma):** A thin membrane present in many peripheral nerves that surrounds the nerve fiber and the myelin sheath

- **Endoneurium:** Loose, or *areolar,* connective tissue surrounding individual fibers

- **Fasciculi:** Bundles of fibers within a nerve

- **Perineurium:** The same kind of connective tissue as endoneurium; surrounds a bundle of fibers

- **Epineurium:** The same kind of connective tissue as endoneurium and perineurium; surrounds several bundles of fibers

There also is a class of cells called *neuroglia,* or simply *glia,* that act as the supportive cells of the nervous system, providing neurons with nutrients and otherwise protecting them. Glia include oligodendrocytes that support the myelin sheath within the central nervous system; star-shaped cells called *astrocytes* that both support nerve tissue and contribute to repairs when needed; and *microgliacytes*, cells that remove dead or dying parts of tissue (this type of cell is called a *phagocyte,* which literally translates from the Greek words for "cell that eats").

Impulses

Neuron membranes are *semi-permeable* (meaning that certain small molecules like ions can move in and out but larger molecules can't), and they're *electrically polarized* (meaning that positively charged ions called *cations* rest around the outside membrane surface while negatively charged ions called *anions* line the inner surface; you can find more about ions in Chapter 1).

A neuron that isn't busy transmitting an impulse is said to be at its *resting potential.* But the *nerve impulse theory,* or *membrane theory,* says that things switch around when a stimulus — a nerve impulse, or *action potential* — moves along the neuron. A stimulus changes the specific permeability of the fiber membrane and causes a depolarization due to a reshuffling of the cations and anions. This change spreads along the nerve fiber and constitutes the nerve impulse. It's called an *all-or-none response* because each neuron has a specific threshold of excitation. Once that threshold is exceeded, the nerve fiber responds with a fixed impulse. After depolarization, repolarization occurs followed by a refractory period, during which no further impulses occur, even if the stimuli's intensity increases.

Intensity of sensation, however, depends on the frequency with which one nerve impulse follows another and the rate at which the impulse travels. That rate is determined by the diameter of the impacted fiber and tends to be more rapid in large nerve fibers. It's also more rapid in myelinated fibers than non-myelinated fibers. The cytoplasm of the axon or nerve fiber is electrically conductive and the myelin decreases the capacitance to prevent charge leakage through the membrane. Depolarization at one node of Ranvier is sufficient to trigger regeneration of the voltage at the next node. Therefore, in myelinated nerve fibers, the action potential does not move as a wave but recurs at successive nodes, traveling faster than in non-myelinated fibers. This is referred to as *saltatory conduction* (from the Latin word *saltare,* which means "to hop or leap").

Synapses

Neurons don't touch, which means that when a nerve impulse reaches the end of a neuron, it needs to cross a gap to the next neuron or to the gland or muscle cell for which the message is intended. That gap is called a *synapse,* or *synaptic cleft.* There are two types of synapses: electrical and chemical. An electric synapse — generally found in organs and between glial cells — uses channels known as *gap junctions* to permit direct transmission of signals between neurons. But in other parts of the body, chemical changes occur to let the impulse cross the gap. The end branches of an axon each form a terminal knob or bulb called a *bouton terminal* (that first word's pronounced *boo-taw*), beyond which there is a space between it and the next nerve pathway. (See the bouton terminal in Figure 15-1.) When an impulse reaches the bouton terminal, the following happens:

1. **Synaptic vesicles in the knob release a transmitter called *acetylcholine* that flows across the gap and increases the permeability of the next cell membrane in the chain.**

2. **An enzyme called *cholinesterase* breaks the transmitter down into *acetyl* and *choline,* which then diffuse back across the gap.**

3. **An enzyme called *choline acetylase* in the synaptic vesicles reunites the acetyl and choline, prepping the bouton terminal to do its job again when the next impulse rolls through.**

A neuron that transmits information toward a synapse is called a *presynaptic neuron.* A neuron that transmits information away from a synapse is called a *postsynaptic neuron.* A neuron can be presynaptic at one synapse and postsynaptic at another. Synapses occur between a presynaptic neuron and a dendrite or cell body of a postsynaptic neuron.

Nervous about getting all this right? Try some practice questions:

1.–5. Match the term to its description.

1. _____ Irritability

2. _____ Conductivity

3. _____ Orientation

4. _____ Coordination

5. _____ Conceptual thought

a. Tissue's ability to respond to stimulation

b. Ability to receive impulses and direct them to channels for favorable response

c. Sense organs' capacity to generate nerve impulse to stimulation

d. Spreading of the nerve impulse

e. Capacity to record, store, and relate information to be used to determine future action

6. What are the three primary functions of the nervous system?

   a. Motion, circulation, conceptual thought

   b. Action, conduction, filtration

   c. Perception, integration, action

   d. Circulation, coordination, response

7. There are many parts in the nervous system, but which is the functional unit?

   a. Axon

   b. Nephron

   c. Dendrite

   d. Neuron

8. Which structure can never be found at the end of a neuron's cytoplasmic projection?

   a. Node of Ranvier

   b. End organ

   c. Effector

   d. End bulb

9. What does the afferent fiber called a dendrite do?

   a. Conducts nerve impulses away from the cell body

   b. Attaches collaterals to the cell body

   c. Conducts nerve impulses to the cell body

   d. Directs neurofibrillae to the proper projection

10. Which membrane surrounds an axon (nerve fiber)?

   a. Sarcolemma

   b. Neurilemma

   c. Perineurium

   d. Epineurium

11.–15. Match the term to its description.

   11. _____ Astrocytes

   12. _____ Microgliacytes

   13. _____ Oligodendrocytes

   14. _____ Axons

   15. _____ Dendrites

   a. Cytoplasmic projections carrying impulses to the cell body

   b. Cells that form and preserve myelin sheaths

   c. Cytoplasmic projections carrying impulses from the cell body

   d. Cells that are phagocytic

   e. Cells that contribute to the repair process of the central nervous system

16. Neuroglia cells provide all of the following to neurons:

   a. Conduction, motor function, nutrients

   b. Support, insulation, nutrients

   c. Sensing function, support, motor function

   d. Stimuli detection, conduction, support

17. How many dendrites and axons are there on a multipolar neuron?

   a. Many dendrites and one axon

   b. Three dendrites and three axons

   c. Four or more dendrites and two axons

   d. A single dendrite and a single axon

18. A synapse between neurons is best described as the transmission of

   a. a continuous impulse.

   b. an impulse through chemical and physical changes.

   c. an impulse through a physical change.

   d. an impulse through a chemical change.

19.–23. Match the term to its description.

   19. _____ Endoneurium   **a.** Fatty layer around an axon fiber

   20. _____ Neurilemma   **b.** Outer thin membrane around an axon fiber

   21. _____ Schwann cell   **c.** Cell in the sheath of an axon

   22. _____ Node of Ranvier   **d.** Depression in the sheath around a fiber

   23. _____ Myelin sheath   **e.** Connective tissue surrounding individual fibers in a nerve

24.–28. Match the term to its description.

   24. _____ All-or-none response   **a.** Impermeability of cell membrane

   25. _____ Cation   **b.** Negatively charged ion on the inner surface of the cell membrane

   26. _____ Anion

   27. _____ Polarization   **c.** Threshold of excitation determines ability to respond

   28. _____ Depolarization   **d.** Positively charged ion on the outer surface of the cell membrane

                      e. Reshuffling of cell membrane ions; permeability of cell membrane

29.–33. Match the term to its description.

29. _____ Cholinesterase

30. _____ Choline acetylase

31. _____ Terminal bulb

32. _____ Acetylcholine

33. _____ Synapse

a. Excitatory chemical necessary for continual nerve pathway

b. Enzyme for breakdown of excitatory chemical

c. Enzyme for reformation of excitatory chemical

d. Space between neurons

e. Contains storage vesicles for excitatory chemical

Minding the Central Nervous System

Together, the brain and the spinal cord make up the central nervous system. You get the scoop on both parts in the following sections.

The spinal cord

The spinal cord has two primary functions: It conducts nerve impulses to and from the brain, and it processes sensory information to produce a spinal reflex without initiating input from the higher brain centers.

The spinal cord, which forms very early in the embryonic spinal canal, extends down into the tail portion of the vertebral column. But because bone grows much faster than nerve tissue, the end of the cord soon is too short to extend into the lowest reaches of the spinal canal. In an adult, the 18-inch spinal cord ends between the first and second lumbar vertebrae, roughly where the last ribs attach. Its tapered end is called the *conus medullaris.* The cord continues as separate strands below that point and is referred to as the *cauda equina* (horse tail). A thread of fibrous tissue called the *filum terminale* extends to the base of the *coccyx* (tailbone) and is attached by the coccygeal ligament.

An oval-shaped cylinder with two deep grooves running its length at the back and the front, the spinal cord doesn't fill the spinal cavity by itself. Also packed inside are the *meninges,* cerebrospinal fluid, a cushion of fat, and various blood vessels.

Three membranes called *meninges* envelop the central nervous system (brain and spinal cord), separating it from the bony cavities.

- ✔ **The dura mater:** The outer layer, the *dura mater,* is the hardest, toughest, and most fibrous layer and is composed of white collagenous and yellow elastic fibers.

- ✔ **The arachnoid:** The middle membrane, the *arachnoid,* forms a weblike layer just inside the dura mater.

- ✔ **The pia mater:** A thin inner membrane, the *pia mater,* lies close along the surface of the central nervous system. The pia mater and arachnoid may adhere to each other and be considered as one, called the *pia-arachnoid.*

There are spaces or cavities between the pia mater and the arachnoid where major regions join, for instance where the medulla oblongata and the cerebellum join. These sub-arachnoid spaces are called *cisterna.* Spaces or cavities between the arachnoid layer and the dura mater layer are referred to as *subdural*. The meninges as well as the ventricles of the brain (described later in this chapter) contain a watery fluid, cerebrospinal fluid (CSF), that surrounds the central nervous system and acts as a cushion.

Two types of solid material make up the inside of the cord, which you can see in Figure 15-2: *gray matter* (which is indeed grayish in color) containing unmyelinated neurons, dendrites, cell bodies, and neuroglia; and *white matter (funiculus),* so-called because of the whitish tint of its myelinated nerve fibers. At the cord's midsection is a small *central canal* filled with CSF and surrounded first by gray matter in the shape of the letter H and then by white matter, which fills in the areas around the H pattern. The legs of the H are called anterior, posterior, and lateral *horns* of gray matter, or *gray columns.*

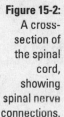

Figure 15-2:
A cross-section of the spinal cord, showing spinal nerve connections.

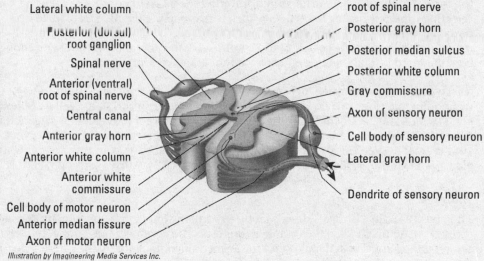

Lateral white column

Posterior (dorsal) root ganglion

Spinal nerve

Anterior (ventral) root of spinal nerve

Central canal

Anterior gray horn

Anterior white column

Anterior white commissure

Cell body of motor neuron

Anterior median fissure

Axon of motor neuron

Posterior (dorsal) root of spinal nerve

Posterior gray horn

Posterior median sulcus

Posterior white column

Gray commissure

Axon of sensory neuron

Cell body of sensory neuron

Lateral gray horn

Dendrite of sensory neuron

Illustration by Imagineering Media Services Inc.

The white matter consists of thousands of myelinated nerve fibers arranged in three *funiculi* (columns) on each side of the spinal cord that convey information up and down the cord's tracts. Ascending afferent (sensory) nerve tracts carry impulses to the brain; descending efferent (motor) nerve tracts carry impulses from the brain. Each tract is named according to its origin and the joint of synapse, such as the *corticospinal tract* (a descending tract with origin at the cortex of the cerebrum and synapsing in the spine) and the *spinothalmic tract* (an ascending tract originating in the spine and synapsing in the thalamus of the brain).

Thirty-one pairs of spinal nerves arise from the sides of the spinal cord and leave the cord through spaces called the *intervertebral foramina* to form the peripheral nervous system, which we discuss in the later section "Taking Side Streets: The Peripheral Nervous System."

The brain

One of the largest organs in the adult human body, the brain tips the scales at 3 pounds and packs roughly 100 billion neurons (yes, that's billion with a "b") and 900 billion supporting neuroglia cells. In this section, we review six major divisions of the brain from the bottom up: *medulla oblongata, pons, midbrain, cerebellum, diencephalon,* and *cerebrum.* We also discuss the ventricles.

The medulla oblongata

The spinal cord meets the brain at the *medulla oblongata,* or lower half of the *brainstem,* just below the right and left cerebellar hemispheres of the brain. In fact, the medulla oblongata is continuous with the spinal cord at its base (inferiorly). It's located anteriorly to the cerebellum and superiorly to the lower border of the pons. All the afferent and efferent tracts of the cord can be found in the brainstem as part of two bulges of white matter forming an area referred to as the *pyramids.* Many of the tracts cross from one side to the other at the pyramids, which explains why the right side of the brain controls the voluntary movement of the left side of the body and vice versa.

Along with the pons, the medulla oblongata also forms a network of gray and white matter called the *reticular formation,* the upper part of the so-called *extrapyramidal pathway.* With its capacity to arouse the brain to wakefulness, it keeps the brain alert, directs messages in the form of impulses, monitors stimuli entering the sense receptors (accepting some and rejecting others it deems to be irrelevant), is responsible for the sleep/wake cycle, refines body movements, and effects higher mental processes such as attention, introspection, and reasoning. Although the cortex of the cerebrum is the actual powerhouse of thought, it must be stimulated into action by signals from the reticular formation.

Nerve cells in the brainstem are grouped together to form nerve centers (nuclei) that control bodily functions, including cardiac activities, and respiration as well as reflex activities such as sneezing, coughing, vomiting, and alimentary tract movements. The medulla oblongata affects these reactions through the vagus, also referred to as cranial nerve X or the 10th cranial nerve. Three other cranial nerves also originate from this area: the 9th (IX) or glossopharyngeal, 11th (XI) or accessory, and 12th (XII) or hypoglossal.

The pons

The *pons* (literally "bridge") does exactly as its name implies: It connects to the cerebellum through structures called the middle peduncles. The cerebellum is connected to the midbrain by the superior peduncles and to the medulla oblongata by the inferior peduncles. It also unites the cerebellar hemispheres, coordinates muscles on both sides of the body, controls facial muscles (including those used to chew), and regulates the first stage of respiration. Oh, and it contains the nuclei for the following cranial nerves: the 5th (V) or trigeminal, 6th (VI) or abducens, 7th (VII) or facial, and 8th (VIII) or vestibulocochlear.

The midbrain

Between the pons and the diencephalon lies the *mesencephalon,* or *midbrain.* It contains the *corpora quadrigemina,* which correlates optical and tactile impulses as well as regulates muscle tone, body posture, and equilibrium through reflex centers in the *superior colliculus.* The *inferior colliculus* contains auditory reflex centers and is believed to be responsible for the detection of musical pitch. The midbrain contains the *cerebral aqueduct (aqueduct of Silvus),* which connects the third ventricle of the diencephalon with the fourth ventricle of the medulla oblongata (see the section "The ventricles" later in this chapter for more). The mesencephalon contains nuclei for the 3rd (III) or oculomotor cranial nerve and the 4th (IV) or trochlear cranial nerve. The red nucleus that contains *fibers* of the *rubrospinal tract,* a motor tract that acts as a relay station for impulses from the cerebellum and higher brain centers, also lies within the midbrain, constituting the superior cerebellar peduncles.

The cerebellum

The *cerebellum* also is known as the *little brain* or *small brain.* The second-largest division of the brain, it's just above and overhangs the medulla oblongata and lies just beneath the rear portion of the cerebrum. Inside, the cerebellum resembles a tree called the *arbor vitae,* or "tree of life." A central body called the *vermis* connects the two lateral masses called the *cerebellar hemispheres* and assists in motor coordination and refinement of muscular movement, aiding equilibrium and muscle tone. The cerebellar cortex or gray matter contains Purkinje neurons that have pear-shaped cell bodies characterized by a large number of dendritic spines and a single long axon. Most Purkinje neurons release a neurotransmitter called GABA (gamma-aminobutyric acid), which exerts inhibitory action on certain neurons, reducing transmission of nerve impulses as well as regulating and coordinating motor movement. These neurons send impulses to the white matter of the cerebellum and to other deeper nuclei in the cerebellum, and then to the brainstem. The cerebellar cortex has parallel ridges called the *folia cerebelli,* which are separated by deep sulci.

The diencephalon

The *diencephalon,* a region between the midbrain and the cerebrum, contains separate brain structures called the *thalamus, epithalamus, subthalamus,* and *hypothalamus.*

- **The thalamus** is a midline symmetrical structure made up of two halves surrounding the third ventricle. The region where the two sides of the thalami come in contact and join forces is called the *intermediate mass.* The thalamus is a primitive receptive center through which the sensory impulses travel on their way to the cerebral cortex. Here, nerve fibers from the spinal cord and lower parts of the brain synapse with neurons leading to the sensory areas of the cortex of the cerebrum. The thalamus is the great integrating center of the brain with the ability to correlate the impulses from tactile, pain, and gustatory (taste) senses with motor reactions.

- **The epithalamus** contains the *choroid plexus,* a vascular structure that produces cerebrospinal fluid. The pineal body and olfactory centers also lie within the epithalamus, which forms the roof of the third ventricle.

- **The subthalamus** is located below the thalamus and regulates the muscles of emotional expression.

- **The hypothalamus** contains the centers for sexual reflexes; body temperature; water, carbohydrate, and fat metabolism; and emotions that affect the heartbeat and blood pressure. It also has the *optic chiasm* (connecting the optic nerves to the optic tract), the posterior lobe of the pituitary gland, and a funnel-shaped region called the *infundibulum* that forms the stalk of the pituitary gland.

The *limbic system* is a complex set of brain structures that lie on both sides of the thalamus. It functions to support certain emotions and behavior, motivation, the formation of long-term memories, and olfaction. Emotions related to the limbic system include fear, anger, and various emotions related to sexual behavior. Two structures in the limbic system, the amygdala and the hippocampus, play important roles in memories. The amygdala is involved in the storage of memories while the hippocampus sends memories to the appropriate parts of the cerebral hemisphere for long-term storage and retrieves them when necessary.

The cerebrum

The *cerebrum,* or *telencephalon,* is often called the *true brain.* It has two cerebral hemispheres — the right and the left. A thin outer layer of gray matter called the *cerebral cortex* features folds or convolutions called *gyri;* furrows and grooves are referred to as *sulci,* and deeper grooves are called *fissures.* A longitudinal fissure separates the cerebral hemispheres. The transverse fissure separates the cerebrum and the cerebellum. Each hemisphere has a set of controls for sensory and motor activities of the body. Interestingly, it's not just right-side/left-side controls that are reversed in the cerebrum; the upper areas of the cerebral cortex control the lower body activities while the lower areas of the cortex control upper-body activities in a reversal called "little man upside down."

Commissural fibers, a tract of nerves running from one side of the brain to the other, coordinate activities between the right and left hemispheres. The *corpus callosum* physically unites the two hemispheres and is the largest and densest mass of commissural fibers. A smaller mass called the *fornix* also plays a role.

Different functional areas of the cerebral cortex are divided into lobes:

- **Frontal lobe:** The seat of intelligence, memory, and idea association

- **Parietal lobe:** Functions in the sensations of temperature, touch, and sense of position and movement as well as the perception of size, shape, and weight

- **Temporal lobe:** Responsible for perception and correlation of acoustical stimuli

- **Occipital lobe:** Handles visual perception

The frontal lobe and the parietal lobe are separated from the temporal lobe below by the *Silvian (lateral) fissure.*

The *medulla,* the region interior to the cortex, is composed of white matter that consists of three groups of fibers:

- **Projection fibers** carry impulses afferently from the brain stem to the cortex and efferently from the cortex to the lower parts of the central nervous system.

- **Association fibers** originate in the cortical cells and carry impulses to the other areas of the cortex on the same hemisphere.

- **Commissural fibers** connect the two cerebral hemispheres.

The ventricles

The brain's four *ventricles* are cavities and canals filled with cerebrospinal fluid. Two lateral ventricles are separated by the *septum pellucidum,* one in each of the cerebral hemispheres. The lateral ventricles communicate with the third ventricle through the *foramen of Monro.* The *cerebral aqueduct* connects the third ventricle to the fourth ventricle, which is continuous with the central canal of the spinal cord and contains openings to the meninges. The fourth ventricle has perforations that allow fluid to enter into the subarachnoid spaces of the meninges. Three foramina allow cerebrospinal fluid to enter the subarachnoid spaces: two foramina of Luschka and one foramen of Magendie.

Lining the ventricles is a thin layer of epithelial cells known as *ependyma,* or the *ependymal layer.* Along with a network of capillaries from the pia mater, the ependyma and capillaries form the *choroid plexus,* which is the source of cerebrospinal fluid. The choroid plexus of each lateral ventricle produces the greatest amount of fluid. Fluid formed by the choroid plexus filters out by *osmosis* (refer to Chapter 2) and circulates through the ventricles. Fluid is returned to the blood through the *arachnoid villi,* fingerlike projections of the *arachnoid meninx,* which absorbs the fluid.

Twelve pairs of cranial nerves connect to the central nervous system via the brain (as opposed to the 31 pairs that connect via the spinal cord). Cranial nerves are identified by Roman numerals I through XII, and memorizing them is a classic test of anatomical knowledge. Check out Table 15-1 for a listing of all the nerves.

Table 15-1		Cranial Nerves	
Number	**Name**	**Type**	**Function**
I	Olfactory	Sensory	Smell
II	Optic	Sensory	Vision
III	Oculomotor	Mixed nerve	Eyeball muscles
IV	Trochlear	Mixed nerve	Eyeball muscles
V	Trigeminal	Tri means "three," so the three types of trigeminal nerves are 1) Opthalmic nerve: sensory nerve; skin and mucous membranes of face and head; 2) Maxillary nerve: mixed nerve; mastication; 3) Mandibular nerve: mixed nerve; mastication	Skin; mastication (chewing)
VI	Abducens	Mixed nerve	Eye movements laterally
VII	Facial	Mixed nerve	Facial expression; salivary secretion; taste
VIII	Vestibulocochlear	Sensory	Auditory nerve for hearing and equilibrium
IX	Glossopharyngeal	Mixed nerve	Taste; swallowing muscles of pharynx; salivation; gag reflex; speech
X	Vagus	Mixed nerve	Controls most internal organs (viscera) from head and neck to transverse colon
XI	Accessory	Mixed nerve	Swallowing and phonation; rotation of head and shoulder
XII	Hypoglossal	Motor nerve	Tongue movements

The first letters of each of these nerve names, in order, are OOOTTAFVGVAH. That's a mouthful, but students have come up with a number of memory tools to remember them. Our favorite is: Old Opera Organs Trill Terrific Arias For Various Grand Victories About History.

Put your knowledge of the central nervous system to the test:

Q. The meninges' functions are primarily

  a. immunological.

  b. supportive.

  c. protective.

  d. a and b.

  e. b and c.

A. The correct answer is b and c (supportive and protective). Yes, meninges have two functions.

34. The cerebrum is divided into two major halves, called _____, by the _____.

   a. cerebellar hemispheres, transverse fissure

   b. cerebral spheres, central sulcus

   c. cerebellar spheres, fissure of Sylvius

   d. cerebral hemispheres, longitudinal fissure

35. You've been in an accident, and you have injured the left side of your cerebrum. Which areas of your body would you most likely have trouble moving?

   a. The right side

   b. The lower body

   c. The upper extremities

   d. The left side

36. Where within the cerebrum does visual activity take place?

   a. Parietal lobe

   b. Medulla oblongata

   c. Occipital lobe

   d. Cerebellum

37. What, among many things, does the cerebellum do?

   a. Controls visual activity

   b. Generates associative reasoning

   c. Interprets auditory stimulation

   d. Coordinates motor control

38.–42. Match the term to its description.

   38. _____ White matter

   39. _____ Reticular formation

   40. _____ Funiculus

   41. _____ Medulla oblongata

   42. _____ Gray matter

   a. Has the capacity to arouse the brain to wakefulness

   b. Myelinated fibers

   c. Bundles of nerve fibers arranged in tracts

   d. Junction of the spinal cord and the brain

   e. Unmyelinated fibers, neuron cell bodies, and neuroglia

43.–47. Match the term to its description.

43. _____ Pons

44. _____ Cerebellum

45. _____ Medulla oblongata

46. _____ Cerebrum

47. _____ Mesencephalon

a. Bridge connecting the medulla oblongata and cerebellum

b. Contains the centers that control cardiac, respiratory, and vasomotor functions

c. Contains the corpora quadrigemina and nuclei for the oculomotor and trochlear nerves

d. Controls motor coordination and refinement of muscular movement

e. Controls sensory and motor activity of the body

48. What does the choroid plexus do?

a. Filters cerebrospinal fluid

b. Absorbs cerebrospinal fluid

c. Produces cerebrospinal fluid

d. Eliminates exhausted cerebrospinal fluid

49. The diencephalon contains the

a. mesencephalon, third ventricle, and hypothalamus.

b. corpus callosum, cerebellum, and pons.

c. thalamus, pituitary gland, and optic chiasm.

d. telencephalon, foramen of Luschka, and medulla oblongata.

50.–61. Use the terms that follow to identify the parts of the brain shown in Figure 15-3.

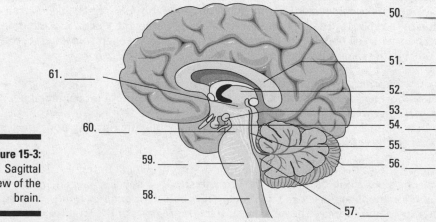

50. _____
51. _____
52. _____
53. _____
54. _____
55. _____
56. _____
57. _____
58. _____
59. _____
60. _____
61. _____

Figure 15-3:
Sagittal
view of the
brain.

© John Wiley & Sons, Inc.

    a. Pons

    b. Thalamus

    c. Cerebellum

    d. Corpus callosum

    e. Third ventricle

    f. Hypothalamus

    g. Cerebrum

    h. Cerebral aqueduct

    i. Midbrain

    j. Pituitary gland

    k. Medulla oblongata

    l. Fourth ventricle

Taking Side Streets: The Peripheral Nervous System

The *peripheral nervous system* is the network that carries information to and from the spinal cord. Among its key structures are 31 pairs of spinal nerves (see Figure 15-4), each originating embryonically in a segment of the spinal cord called a *neuromere.* Eight of the spinal nerve pairs are *cervical* (having to do with the neck), 12 are *thoracic* (relating to the chest, or thorax), five are *lumbar* (between the lowest ribs and the pelvis), five are *sacral* (the posterior section of the pelvis), and one is *coccygeal* (relating to the tailbone). Spinal nerves connect with the spinal cord by two bundles of nerve fibers, or *roots.* The *dorsal root* contains afferent fibers that carry sensory information from receptors to the central nervous system. The cell bodies of these sensory neurons lie outside the spinal cord in a bulging area called the *dorsal root ganglion* (refer to the cross-section of the spinal cord in Figure 15-2). A second bundle, the *ventral root,* contains efferent motor fibers with cell bodies that lie inside the spinal cord. In each spinal nerve, the two roots join outside the spinal cord to form what's called a *mixed spinal nerve.*

Spinal reflexes, or *reflex arcs,* occur when a sensory neuron transmits a "danger" signal — like a sensation of burning heat — through the dorsal root ganglion. An internuncial neuron (or association neuron) in the spinal cord passes along the signal to a motor neuron (or efferent fiber) that stimulates a muscle, which immediately pulls the burning body part away from heat (see Figure 15-5).

After a spinal nerve leaves the spinal column, it divides into two small branches. The posterior, or *dorsal ramus,* goes along the back of the body to supply a specific segment of the skin, bones, joints, and longitudinal muscles of the back. The ventral, or *anterior ramus,* is larger than the dorsal ramus and supplies the anterior and lateral regions of the trunk and limbs.

Groups of spinal nerves interconnect to form an extensive network called a *plexus* (Latin for "braid"), each of which connects through the anterior ramus, including the *cervical plexus* of the neck, *brachial plexus* of the shoulder and axilla, and *lumbosacral plexus* of the lower back (including the body's largest nerve, the *sciatic nerve*). However, there's no plexus in the thoracic area. Instead, the anterior ramus directly supplies the intercostal muscles (literally "between the ribs") and the skin of the region.

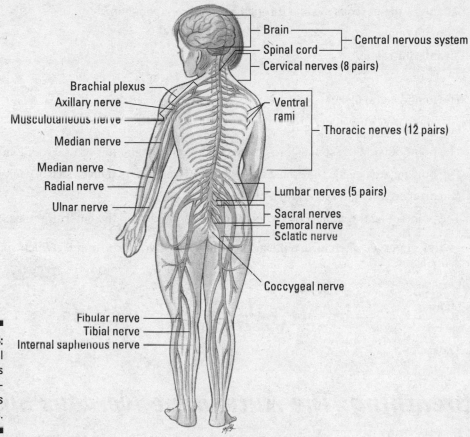

Figure 15-4:
The spinal
nerves plus
branch-
ing plexus
nerves.

Illustration by Kathryn Born, MA

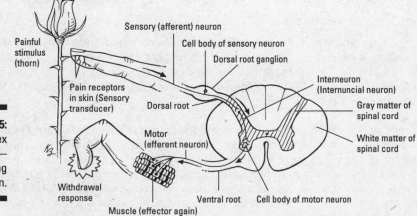

Figure 15-5:
A reflex
arc —
responding
to pain.

Illustration by Kathryn Born, MA

62. What enters the nervous system through a dorsal root ganglion?

    a. A dorsal ramus

    b. An anterior ramus

    c. A reflex arc

    d. The lumbosacral plexus

63. What is a mixed spinal nerve?

    a. The point outside the spinal cord where the dorsal root and the ventral root join

    b. The bulging area inside the spinal cord where afferent fibers deliver sensory information

    c. A neuromere that interacts with a plexus

    d. The point along the spinal cord where the dorsal ramus and the anterior ramus join

64. Networks of nerves often form a plexus, from the Latin word for "braid." Which spinal nerve area does not?

    a. Cervical

    b. Lumbar

    c. Sacral

    d. Thoracic

Keep Breathing: The Autonomic Nervous System

Just as the name implies, the autonomic nervous system functions automatically. Divided into the *sympathetic* and *parasympathetic* systems, it activates the involuntary smooth and cardiac muscles and glands to serve such vital systems that function automatically as the digestive tract, circulatory system, respiratory, urinary, and endocrine systems. Autonomic functions are under the control of the hypothalamus, cerebral cortex, and medulla oblongata.

The sympathetic system, which is responsible for the body's involuntary fight-or-flight response to stress, is defined by the autonomic fibers that exit the thoracic and lumbar segments of the spinal cord. The parasympathetic system is defined by the autonomic fibers that either exit the brainstem via the cranial nerves or exit the sacral segments of the spinal cord. (You can see both systems in Figure 15-6.)

The sympathetic and parasympathetic systems oppose each other in function, helping to maintain *homeostasis,* or balanced activity in the body systems. Yet, often the sympathetic and parasympathetic systems work in concert. The sympathetic system dilates the eye's pupil, but the parasympathetic system contracts it again. The sympathetic system quickens and strengthens the heart while the parasympathetic slows the heart's action. The sympathetic system contracts blood vessels in the skin so more blood goes to muscles for a fight-or-flight reaction to stress, and the parasympathetic system dilates the blood vessels when the stress concludes.

A pair of sympathetic trunks lies to the right and left of the spinal cord and is composed of a series of ganglia that form nodular cords extending from the base of the skull to the front of the coccyx (tailbone). Sympathetic nerves originate as a short preganglionic neuron with its cell body inside the lateral horn of the gray matter of the spinal cord from the first thoracic to the second lumbar. Axons of these nerves then pass through the ventral root of the spinal nerve, leaving it through a branch of the spinal nerve called the *white rami* (named for their white myelin sheaths), which connect to one of the two chains of ganglia in the trunks. Within these hubs, synapses distribute the nerves to various parts of the body.

The parasympathetic system is referred to as a *craniosacral system* because its nuclei originate in the medulla oblongata, pons (brainstem), and the mesencephalon (midbrain), sending out impulses through the oculomotor III nerve, the facial VII nerve, the glossopharyngeal IX nerve, and the vagus X nerve, as well as the nerves from the sacral portion of the spinal cord (refer to Table 15-1). Parasympathetic nerves consist of long, preganglionic fibers that synapse in a terminal ganglion near or within the organ or tissue that's being innervated. Generally speaking, the parasympathetic system acts in opposition to the sympathetic system.

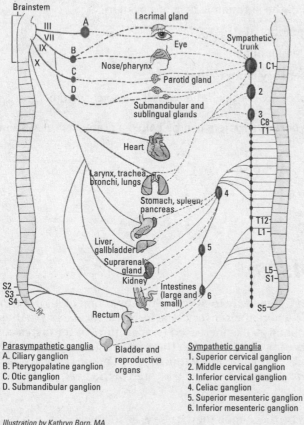

Figure 15-6: The sympathetic and parasympathetic nervous systems.

Parasympathetic ganglia
A. Ciliary ganglion
B. Pterygopalatine ganglion
C. Otic ganglion
D. Submandibular ganglion

Sympathetic ganglia
1. Superior cervical ganglion
2. Middle cervical ganglion
3. Inferior cervical ganglion
4. Celiac ganglion
5. Superior mesenteric ganglion
6. Inferior mesenteric ganglion

Illustration by Kathryn Born, MA

See whether any of the following practice questions touch a nerve:

65. Which three parts of the brain control autonomic functions?

    a. Medulla oblongata, parietal lobe, and cerebellum

    b. Occipital lobe, cerebral cortex, and temporal lobe

    c. Hypothalamus, cerebral cortex, and brainstem

    d. Brainstem, Broca's area, and frontal lobe

66. Which of the following statements is true about the autonomic nervous system?

    a. It has two parts: the parasympathetic that controls all normal functions, and the sympathetic that carries out the same functions.

    b. It's the nervous system that controls all reflexes.

    c. It doesn't function when the body's under stress.

    d. It has two divisions that are antagonistic to each other, meaning that one counteracts the effects of the other one.

67. What is the primary function of the autonomic nervous system?

    a. It activates in response to emotional stress.

    b. It controls involuntary body functions.

    c. It's the primary source of sensory input to the body.

    d. It works to block signals that may overload the overall nervous system.

68. Which part of the autonomic nervous system can be called a craniosacral system?

    a. Ganglia

    b. Sympathetic trunks

    c. Parasympathetic system

    d. Medulla oblongata

Coming to Your Senses

The nervous system must have some way to perceive its environment in order to generate appropriate responses. That's where the senses come in. *Sense receptors* are those numerous organs that respond to stimuli — like increased temperature, bitter tastes, and sharp points — by generating a nerve impulse. Although there are millions of *general sense receptors* found throughout the body that can convey touch, pain, and physical contact, there are far fewer of the *special sense receptors* — those located in the head — that really bring meaning to your world.

Sense receptors are classified by the stimuli they receive, as follows:

    ✔ **Exteroceptors:** Receive stimuli from the external environment. These are sensory nerve terminals, such as those in the skin and mucous membranes, that are stimulated by the immediate external environment.

- **Interoceptors:** Receive stimuli from the internal environment. These can be any of the sensory nerve terminals located in and transmitting impulses from the viscera.

- **Proprioceptors:** Part of the "true" internal environment. They're sensory nerve terminals chiefly found in muscles, joints, and tendons that give information concerning movements and position of the body.

- **Teleceptors:** Sensory nerve terminals stimulated by emanations from distant objects. They exist in the eyes, ears, and nose.

On watch: The eyes

Although there are many romantic notions about eyes, the truth is that an eyeball is simply a hollow sphere bounded by a trilayer wall and filled with a gelatinous fluid called, oddly enough, *vitreous humor.* The outer fibrous coat is made up of the *sclera* (the white of the eye) in back and the *cornea* in front. The sclera provides mechanical support, protection, and a base for attachment of eye muscles, which assist in the focusing process. The cornea covers the anterior with a clear window.

An intermediate, or vascular, coat called the *uvea* provides blood and lymphatic fluids to the eye; regulates the amount of light entering the eye; and secretes and reabsorbs *aqueous humor,* a thin, watery liquid that fills the anterior chamber of the eyeball in front of the iris and the posterior chamber between the iris and the lens. A pigmented coat has three components:

- **The iris:** Contains blood vessels, pigment cells, and smooth muscle fibers to control the pupil's diameter.

- **The ciliary body:** Attached to the periphery of the iris and lens.

- **The choroid:** A thin, dark brown, vascular layer lining most of the sclera on the back and sides of the eye. The choroid contains arteries, veins, and capillaries that supply the *retina* and sclera with nutrients, and it also contains pigment cells to absorb light and prevent reflection and blurring.

The *retina* is part of an internal nervous layer connecting with the optic nerve that leaves at the back (posterior) of each eye. The nervous tissue layers along the inner back of the eye contain *rods* and *cones* (types of neurons that analyze visual input). The rods are dim light receptors whereas the cones detect bright light and construct form, structure, and color. The retina has an *optic disc,* which is essentially a blind spot incapable of producing an image. The eye's blind spot is a result of the absence of photoreceptors in the area of the retina where the optic nerve leaves the eye.

The macula or *macula lutea* (Latin for "yellow spot") is an oval-shaped, highly pigmented yellow area near the center of the retina. About 6 millimeters in diameter, it's defined by two or more layers of ganglion cells. There is a small dimple near the center of the macula called the *fovea centralis* (*fovea* is Latin for "pit"). This is where the eye's vision is sharpest and it is where most of the eye's color perception occurs. Only cones can be found in this area, smaller and more closely packed than anywhere else in the retina. The fovea sees only the central 2 degrees of the visual field, which encompasses roughly twice the width of your thumbnail at arm's length. If an observed area or object is larger than this angle of vision, such as when reading, the eyes must constantly shift their gaze, a process known as *saccadic movement,* to bring different portions of the image into the fovea. For this reason, the cognitive brain is responsible for oculomotor commands and piecing together a comprehensive visual image. Damage to the macula causes the loss of central vision. The progressive destruction of the macula occurs in a disease known as *macular degeneration.*

The crystalline lens consists of concentric layers of protein. It's *biconvex* in shape, bulging outward. Located behind the pupil and iris, the lens is held in place by ligaments attached to the ciliary body, which contains ciliary muscles. When the ciliary muscles contract, the shape of the lens changes, altering the visual focus. This process of *accommodation* allows the eye to see objects both at a distance and close-up.

The *palpebrae* (eyelids) extend from the edges of the eye orbit, into which roughly five-sixths of the eyeball is recessed. Eyelids come together at medial and lateral angles of the eye that are called the *canthi*. In the medial angle of the eye is a pink region called the *caruncula,* or *caruncle*. The caruncula contains sebaceous glands and sudoriferous (sweat) glands. A mucous membrane called the *conjunctiva* covers the inner surface of each eyelid and the anterior surface of the eye, but not the cornea. Up top and to the side of the orbital cavities are lacrimal glands that secrete tears that are carried through a series of lacrimal ducts to the conjunctiva of the upper eyelid. Ultimately, secretions drain from the eyes through the nasolacrimal ducts.

Listen up: The ears

Human ears — otherwise called *vestibulocochlear* organs — are more than just organs of hearing. They also serve as organs of equilibrium, or balance. Here are the three divisions of the ear:

- ✔ **The external ear** includes the *auricle,* or *pinna,* which is the folded, rounded appendage made of elastic cartilage and skin. Extending into the skull is the *ear canal,* or *external auditory meatus,* a short passage through the temporal bone ending at the *tympanic membrane,* or *eardrum*. Sebaceous glands near the external opening and *ceruminous* glands in the upper wall produce the brownish substance known as earwax, or *cerumen*.

- ✔ **The middle ear** is a small, usually air-filled cavity in the skull that's lined with mucous membrane. It communicates through the *Eustachian tube* with the pharynx. The Eustachian tube keeps air pressure equal on both sides of the eardrum (tympanic membrane), equalizing pressure in the middle ear with atmospheric pressure from outside. Three small bones called *auditory ossicles* occupy the middle ear, deriving their names from their shapes: the *malleus* (hammer), the *incus* (anvil), and the *stapes* (stirrup).

- ✔ **The internal ear** is the most complex structure of the entire organ because it's where vibrations are translated. It's composed of a group of interconnected canals or channels called the *cochlea*. Within the cochlea are three canals separated from each other by thin membranes; two of the canals — the *vestibular* and the *tympanic* — are bony chambers filled with a *perilymph fluid,* and the third canal — the *cochlear* canal — is a membranous chamber filled with *endolymph*. The cochlear canal lies between the vestibular and tympanic canals and contains the *organ of Corti,* a spiral-shaped organ made up of cells with projecting hairs that transmit auditory impulses.

The process of hearing a sound follows these basic steps:

1. **Sound waves travel through the auditory canal, striking the eardrum, making it vibrate, and setting the three ossicle bones into motion.**

2. **The stapes at the end of the chain vibrates the oval window of the vestibular canal, translating the motion into the perilymph fluid in the vestibular and tympanic canals of the cochlea. The round window, a membranous partition, separates the perilymph of the tympanic canal from the air spaces of the middle ear. The round window moves in the opposite phase to the oval window, allowing the fluid to move in the cochlea.**

3. **The vibrating fluid begins moving the basilar membrane that separates the two canals, stimulating the endolymph fluid in the membranous area of the cochlea.**

4. **The stimulated endolymph fluid in turn stimulates the hair cells of the organ of Corti, which transmit the impulses to the brain over the auditory (cochlear) nerve, a division of the vestibulocochlear nerve (cranial nerve VIII).**

That's the hearing part of your ears. Equilibrium requires that some additional parts come into play. Three semicircular canals, each with an *ampulla* (or small, dilated portion) at each end, lie at right angles to each other. The *ampullae* connect to a fluid-filled sac called a *utricle,* which in turn connects to another fluid-filled sac called a *saccule.* Both sacs contain regions called *maculae* that are lined with sensitive hairs and contain concretions (solid masses) of calcium carbonate called *otoliths* (or *otoconia*). When linear acceleration pulls at them, the otoliths press on the hair cells and initiate an impulse to the brain through basal sensory nerve fibers. When the head changes position, it causes a change in the direction of force on the hairs. Movement of the hairs stimulates dendrites of the *vestibulocochlear nerve* (the eighth cranial nerve) to carry impulses to the brain.

Q. The most sensitive region of the retina producing the greatest visual acuity is the

   a. blind spot.

   b. cornea.

   c. fovea centralis.

   d. macula lutea.

A. The correct answer is fovea centralis. It's loaded with light-sensitive cones.

69.–82. Use the terms that follow to identify the internal structures of the eye shown in Figure 15-7.

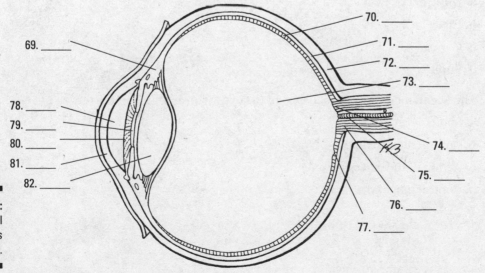

Figure 15-7: The internal structures of the eye.

Illustration by Kathryn Born, MA

a. Blind spot

b. Pupil

c. Optic nerve

d. Retina

e. Cornea

f. Sclera

g. Ciliary body

h. Fovea centralis

i. Lens

j. Anterior chamber (contains aqueous humor)

k. Choroid

l. Blood vessels

m. Iris

n. Posterior cavity (contains vitreous humor)

83. Choose the correct pathway of light through the eye to the retina:

a. Cornea, anterior cavity, pupil, posterior cavity, lens, posterior chamber, retina

b. Cornea, anterior chamber, pupil, posterior chamber, lens, posterior cavity, retina

c. Cornea, pupil, anterior cavity, posterior cavity, posterior chamber, lens, retina

d. Cornea, anterior chamber, iris, pupil, posterior chamber, lens, posterior cavity, retina

84. Which of the following structures is not part of the eyeball?

a. Optic nerve

b. Iris

c. Cornea

d. Pupil

85. The accommodation, or focusing, of the eye involves which of these?

a. Sphincter of the pupil

b. Contraction of the iris

c. Action of the ciliary muscles

d. Contraction of the pupil

86. All the ganglion cells of the retina merge and exit the eye forming the optic nerve. What is the area where they exit called?

   a. Macula lutea

   b. Optic disc

   c. Choroid

   d. Fovea

87. Sound waves travel through the ear canal and cause the tympanic membrane, or eardrum, to vibrate. Those vibrations then are picked up by three small bones, called _____, in the following order: _____.

   a. ossicles; malleus, incus, stapes

   b. tympani; incus, diaphragm, ossicle

   c. ossicles; stapes, malleus, incus

   d. vestibuli; stapedius, tympanus, incus

88. What is the function of the receptor cells in the organ of Corti?

   a. Determining relative position of the body

   b. Receiving impulses from the viscera

   c. Pain perception

   d. Sound perception

89. The vestibular and tympanic canals of the cochlea contain _____ fluid, which carries vibrations that stimulate the _____ fluid in the membranous area of the cochlea.

   a. aqueous humor; vitreous humor

   b. plasma; mucosa

   c. perilymph; endolymph

   d. mastoid; basilar

90. What do utricles and saccules do?

   a. Transmit auditory impulses to the brain

   b. Maintain equilibrium

   c. Vibrate in response to motions of the ossicle bones

   d. Separate the vestibular and tympanic canals

91. What is the difference between aqueous humor and vitreous humor?

   a. One lines the sclera, and the other bathes the iris.

   b. One is thin and watery, and the other is thick and gelatinous.

   c. One acts as a photoreceptor, and the other acts as a focusing mechanism.

   d. One is subtly funny, and the other is brashly obvious.

92.–104. Use the terms that follow to identify the structures of the ear shown in Figure 15-8.

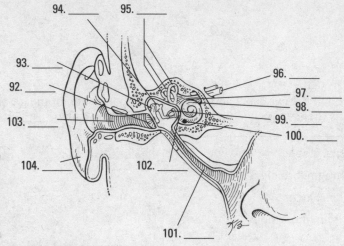

94. _____   95. _____

93. _____

92. _____

103. _____

104. _____   102. _____

101. _____

96. _____

97. _____

98. _____

99. _____

100. _____

Figure 15-8:
The anat-
omy of the
ear.

Illustration by Kathryn Born, MA

   a. Oval window

   b. Semicircular canals

   c. Cochlea

   d. Pinna

   e. Malleus

   f. Cochlear nerve

   g. Stapes

   h. Incus

   i. Round window

   j. External auditory meatus

   k. Tympanic membrane

   l. Vestibulocochlear nerve

   m. Eustachian tube

Answers to Questions on the Nervous System

The following are answers to the practice questions presented in this chapter.

1. Irritability: **a. Tissue's ability to respond to stimulation.**

2. Conductivity: **d. Spreading of the nerve impulse.**

3. Orientation: **c. Sense organs' capacity to generate nerve impulse to stimulation.**

4. Coordination: **b. Ability to receive impulses and direct them to channels for favorable response.**

5. Conceptual thought: **e. Capacity to record, store, and relate information to be used to determine future action.**

6. What are the three primary functions of the nervous system? **c. Perception, integration, action.** Each of the other answer options includes at least one function that clearly belongs to another of the body's systems. (Just so you know, another way to phrase this answer would be orientation, coordination, conceptual thought.)

7. There are many parts in the nervous system, but which is the functional unit? **d. Neuron.** Some of the other answer options are *parts* of a neuron, but the neuron is the central unit.

8. Which structure can never be found at the end of a neuron's cytoplasmic projection? **a. Node of Ranvier.** The nodes of Ranvier are gaps along the myelin sheath, so one of them can't be found at the end of the line.

9. What does the afferent fiber called a dendrite do? **c. Conducts nerve impulses to the cell body.** We gave you a big hint when we referred to it as an afferent fiber. Remember: Afferent connections arrive, and efferent connections exit. Dendrites deliver impulses while axons send them away.

10. Which membrane surrounds an axon (nerve fiber)? **b. Neurilemma.** It actually wraps around the myelin sheath, so it's always on the outside.

11. Astrocytes: **e. Cells that contribute to the repair process of the central nervous system.**

12. Microgliacytes: **d. Cells that are phagocytic.**

13. Oligodendrocytes: **b. Cells that form and preserve myelin sheaths.**

14. Axons: **c. Cytoplasmic projections carrying impulses from the cell body.**

15. Dendrites: **a. Cytoplasmic projections carrying impulses to the cell body.**

16. Neuroglia cells provide all of the following to neurons: **b. Support, insulation, nutrients.**

17. How many dendrites and axons are there on a multipolar neuron? **a. Many dendrites and one axon.** A neuron may have one, many, or no dendrites, but it always has a single axon.

18. A synapse between neurons is best described as the transmission of **d. an impulse through a chemical change.** With all that acetylcholine and cholinesterase floating around, it must be a chemical transmission.

19. Endoneurium: **e. Connective tissue surrounding individual fibers in a nerve.**

20. Neurilemma: **b. Outer thin membrane around an axon fiber.**

21 Schwann cell: **c. Cell in the sheath of an axon.**

22 Node of Ranvier: **d. Depression in the sheath around a fiber.**

23 Myelin sheath: **a. Fatty layer around an axon fiber.**

24 All-or-none response: **c. Threshold of excitation determines ability to respond.**

25 Cation: **d. Positively charged ion on the outer surface of the cell membrane.**

26 Anion: **b. Negatively charged ion on the inner surface of the cell membrane.**

27 Polarization: **a. Impermeability of cell membrane.**

28 Depolarization: **e. Reshuffling of cell membrane ions; permeability of cell membrane.**

29 Cholinesterase: **b. Enzyme for breakdown of excitatory chemical.**

30 Choline acetylase: **c. Enzyme for reformation of excitatory chemical.**

31 Terminal bulb: **e. Contains storage vesicles for excitatory chemical.**

32 Acetylcholine: **a. Excitatory chemical necessary for continual nerve pathway.**

33 Synapse: **d. Space between neurons.**

34 The cerebrum is divided into two major halves, called **d. cerebral hemispheres,** by the **d. longitudinal fissure.** Cerebrum = cerebral, and two halves = hemispheres. You can remember the fissure's name by equating it to Earth's prime meridian, which separates the Eastern and Western Hemispheres, and longitudinal is the most likely position for an equal division.

35 You've been in an accident, and you have injured the left side of your cerebrum. Which areas of your body would you most likely have trouble moving? **a. The right side.** In the cerebrum, right = left and up = down. Clear as mud?

36 Where within the cerebrum does visual activity take place? **c. Occipital lobe.** To remember, use the word "occipital" to bring to mind the word "optic," which of course is related to visual activity.

37 What, among many things, does the cerebellum do? **d. Coordinates motor control.** As the second-largest division of the brain, the cerebellum also is known as the "little brain."

38 White matter: **b. Myelinated fibers.**

39 Reticular formation: **a. Has the capacity to arouse the brain to wakefulness.**

40 Funiculus: **c. Bundles of nerve fibers arranged in tracts.**

41 Medulla oblongata: **d. Junction of the spinal cord and the brain.**

42 Gray matter: **e. Unmyelinated fibers, neuron cell bodies, and neuroglia.**

43 Pons: **a. Bridge connecting the medulla oblongata and cerebellum.**

44 Cerebellum: **d. Controls motor coordination and refinement of muscular movement.**

45 Medulla oblongata: **b. Contains the centers that control cardiac, respiratory, and vasomotor functions.**

46 Cerebrum: **e. Controls sensory and motor activity of the body.**

47 Mesencephalon: **c. Contains the corpora quadrigemina and nuclei for the oculomotor and trochlear nerves.**

48 What does the choroid plexus do? **c. Produces cerebrospinal fluid.** This one requires rote memorization — sorry!

49 The diencephalon contains the **c. thalamus, pituitary gland, and optic chiasm.** Think of it as the home of the thalamus, and you can't go wrong.

50–61 Following is how Figure 15-3, the brain, should be labeled.

50. **g. Cerebrum;** 51. **d. Corpus callosum;** 52. **b. Thalamus;** 53. **f. Hypothalamus;** 54. **j. Pituitary gland;** 55. **h. Cerebral aqueduct;** 56. **c. Cerebellum;** 57. **l. Fourth ventricle;** 58. **k. Medulla oblongata;** 59. **a. Pons;** 60. **i. Midbrain;** 61. **e. Third ventricle.**

62 What enters the nervous system through a dorsal root ganglion? **c. A reflex arc.** That's the sudden sensation of pain or burning heat, also known as a spinal reflex.

63 What is a mixed spinal nerve? **a. The point outside the spinal cord where the dorsal root and the ventral root join.** That juncture occurs in each of the 31 pairs of spinal nerves.

64 Networks of nerves often form a plexus, from the Latin word for "braid." Which spinal nerve area does not? **d. Thoracic.** There is far less nerve networking to be "braided" there.

65 Which three parts of the brain control autonomic functions? **c. Hypothalamus, cerebral cortex, and brainstem.** We may have thrown you for a loop by calling the medulla oblongata and the pons by their more common name: brainstem.

66 Which of the following statements is true about the autonomic nervous system? **d. It has two divisions that are antagonistic to each other, meaning that one counteracts the effects of the other one.** As a result, the body achieves homeostasis.

67 What is the primary function of the autonomic nervous system? **b. It controls involuntary body functions.** That's the only answer option with a sense of automation.

68 Which part of the autonomic nervous system can be called a craniosacral system? **c. Parasympathetic system.** It originates in both the brainstem and the sacral region.

69–82 Following is how Figure 15-7, the internal structures of the eye, should be labeled.

69. **g. Ciliary body;** 70. **d. Retina;** 71. **k. Choroid;** 72. **f. Sclera;** 73. **n. Posterior cavity (filled with vitreous humor);** 74. **a. Blind spot;** 75. **l. Blood vessels;** 76. **c. Optic nerve;** 77. **h. Fovea centralis;** 78. **j. Anterior chamber (contains aqueous humor);** 79. **b. Pupil;** 80. **m. Iris;** 81. **e. Cornea;** 82. **i. Lens.**

83 Choose the correct pathway of light through the eye to the retina: **b. Cornea, anterior chamber, pupil, posterior chamber, lens, posterior cavity, retina.**

You may want to develop your own mnemonic for this path. Here's an unwieldy example: Continuous Air Conditioning Puts Possible Charges Lower on the Politically Correct Radar.

84 Which of the following structures is not part of the eyeball? **a. Optic nerve.** This nerve carries the visual signals to the brain.

85 The accommodation, or focusing, of the eye involves which of these? **c. Action of the ciliary muscles.** They reshape the lens by contracting and relaxing as needed to bring things into focus.

86 All the ganglion cells of the retina merge and exit the eye forming the optic nerve. What is the area where they exit called? **b. Optic disc.** Sometimes referred to as the *optic nerve head,* the optic disc has no photoreceptors of its own, causing a "blind spot" in each of your eyes.

87 Sound waves travel through the ear canal and cause the tympanic membrane, or eardrum, to vibrate. Those vibrations then are picked up by three small bones, called **a. ossicles,** in the following order: **a. malleus, incus, stapes.**

A mnemonic can be helpful here: Mailing Includes Stamps.

88 What is the function of the receptor cells in the organ of Corti? **d. Sound perception.** Hairs in this structure ultimately send the signal down the auditory nerve.

89 The vestibular and tympanic canals of the cochlea contain **c. perilymph** fluid, which carries vibrations that stimulate the **c. endolymph** fluid in the membranous area of the cochlea. Don't forget that the prefix *peri–* means "around" or "near" and the prefix *endo–* means "within."

90 What do utricles and saccules do? **b. Maintain equilibrium.** These little endolymph-filled sacs have hairs and chunks of calcium carbonate that detect changes in gravitational forces.

91 What is the difference between aqueous humor and vitreous humor? **b. One is thin and watery, and the other is thick and gelatinous.** Vitreous humor (which is thick and gelatinous) fills the hollow sphere of the eyeball, but aqueous humor (which is thin and watery) fills two chambers toward the front of the eyeball.

92–**104** Following is how Figure 15-8, the structures of the ear, should be labeled.

92. **k. Tympanic membrane;** 93. **e. Malleus;** 94. **h. Incus;** 95. **b. Semicircular canals;** 96. **l. Vestibulocochlear nerve;** 97. **f. Cochlear nerve;** 98. **c. Cochlea;** 99. **g. Stapes;** 100. **i. Round window;** 101. **m. Eustachian tube;** 102. **a. Oval window;** 103. **j. External auditory meatus;** 104. **d. Pinna.**

Chapter 16

Raging Hormones: The Endocrine System

· ·

In This Chapter

▶ Absorbing what endocrine glands do

▶ Checking out the ringmasters: The pituitary and hypothalamus glands

▶ Surveying the supporting glands

▶ Understanding how the body balances under stress

· ·

The human body has two separate command and control systems that work in harmony most of the time but also work in very different ways. Designed for instant response, the nervous system cracks its cellular whip using electrical signals that make entire systems hop to their tasks with no delay (you can get a feel for the nervous system in Chapter 15). By contrast, the endocrine system's glands use chemical signals called *hormones* that behave like the steering mechanism on a large, fully loaded ocean tanker; small changes can have big impacts, but it takes quite a bit of time for any evidence of the change to make itself known. At times, parts of the nervous system stimulate or inhibit the secretion of hormones, and some hormones are capable of stimulating or inhibiting the flow of nerve impulses.

The word "hormone" originates from the Greek word *hormao*, which literally translates as "I excite." And that's exactly what hormones do. Each chemical signal stimulates specific parts of the body, known as *target tissues* or *target cells*. The body needs a constant supply of hormonal signals to grow, maintain homeostasis, reproduce, and conduct myriad processes.

In this chapter, we go over which glands do what and where, as well as review the types of chemical signals that play various roles in the body. You also get to practice discerning what the endocrine system does, how it does it, and why the body responds like it does.

No Bland Glands

Technically, there are ten or so primary endocrine glands, made up of epithelial cells embedded within connective tissue, with various other hormone-secreting tissues scattered throughout the body. Unlike *exocrine glands* (such as mammary glands and sweat glands), endocrine glands have no ducts to convey their secretions. Instead, hormones move directly into extracellular spaces surrounding each endocrine gland and from there move into capillaries and the greater bloodstream. Although they spread throughout the body in the bloodstream, hormones are uniquely tagged by their chemical composition. Thus they have separate identities and stimulate specific receptors on target cells so that usually only the intended cells or tissues respond to their signals. Water-soluble hormones affect a target cell by binding to receptors on the cell's membrane, while lipid-soluble hormones diffuse into the cell.

All of the many hormones can be classified either as *steroid* (derived from cholesterol) or *nonsteroid* (derived from amino acids and other proteins). The steroid hormones — which include testosterone, estrogen, progesterone, and cortisol — are the ones most closely associated with emotional outbursts and mood swings. Because they are lipid-soluble, steroidal hormones, which are nonpolar (see Chapter 2 for details on cell diffusion), they penetrate cell membranes easily and initiate protein production at the nucleus.

Nonsteroid hormones are divided among four classifications:

- ✔ Some are derived from *modified amino acids,* including such things as epinephrine and norepinephrine, as well as melatonin.

- ✔ Others are *peptide*-based, including an antidiuretic hormone called ADH, oxytocin, and a melanocyte-stimulating hormone called MSH.

- ✔ *Glycoprotein*-based hormones include follicle-stimulating hormone (FSH), luteinizing hormone (LH), and chorionic gonadotropin — all closely associated with the reproductive systems (which we cover in Chapters 13 and 14).

- ✔ *Protein*-based nonsteroid hormones include such crucial substances as insulin and growth hormone as well as prolactin and parathyroid hormone.

Hormone functions include controlling the body's internal environment by regulating its chemical composition and volume, activating responses to changes in environmental conditions to help the body cope, influencing growth and development, enabling several key steps in reproduction, regulating components of the immune system, and regulating organic metabolism.

See if all this hormone-speak is sinking in:

1.–5. Mark the statement with a T if it's true or an F if it's false:

1. _____ The endocrine system brings about changes in the metabolic activities of the body tissue.

2. _____ The amount of hormone released is determined by the body's need for that hormone at the time.

3. _____ The glands of the endocrine system are composed of cartilage cells.

4. _____ Endocrine glands aren't functional in reproductive processes.

5. _____ Some hormones can be derivatives of amino acids, whereas others are synthesized from cholesterol.

6. _____ glands secrete their product through ducts while _____ glands secrete their product into the interstitial fluid, which flows into the blood.

a. Exocrine; endocrine

b. Endocrine; exocrine

c. Heterocrine; interocrine

d. Pericrintal; heterocrine

7. Some hormones affect a cell by binding to membrane receptors, while others diffuse into the cell. Hormones that bind to membrane receptors are called _____, and hormones that diffuse into the cell are called _____.

a. lipid-soluble; cyclic

b. water-soluble; lipid-soluble

c. lipid-soluble; water-soluble

d. cyclic; water-soluble

Enter the Ringmasters

The key glands of the endocrine system include the *pituitary* (also called the *hypophysis*), *adrenal* (also referred to as *suprarenal*), *thyroid, parathyroid, thymus, pineal, islets of Langerhans* (within the *pancreas*), and *gonads* (testes in the male and ovaries in the female). But of all these, it's the pituitary working in concert with the hypothalamus in the brain that really keeps things rolling (see Figure 16-1).

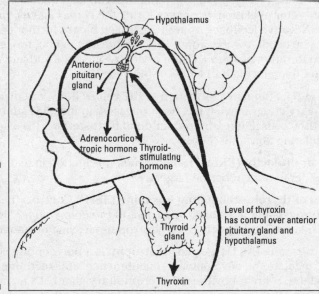

Figure 16-1: The working relationship of the hypothalamus and the pituitary gland.

Illustration by Kathryn Born, MA

Labels in figure: Hypothalamus; Anterior pituitary gland; Adrenocorticotropic hormone; Thyroid-stimulating hormone; Thyroid gland; Thyroxin; Level of thyroxin has control over anterior pituitary gland and hypothalamus

The hypothalamus

REMEMBER

The hypothalamus is the unsung hero linking the body's two primary control systems — the endocrine system and the nervous system. Part of the brain and part of the endocrine system, the hypothalamus is connected to the pituitary via a narrow stalk called the *infundibulum* that carries regular system status reports to the pituitary. In its supervisory role, the hypothalamus provides neurohormones to control the pituitary gland and influences food and fluid intake as well as weight control, body heat, and the sleep cycle. Releasing and inhibiting hormones produced by the hypothalamus regulate the release of most anterior pituitary hormones.

The hypothalamus sits just above the pituitary gland, which is nestled in the middle of the human head in a depression of the skull's sphenoid bone called the *sella turcica.*

The pituitary

The pituitary's *anterior lobe,* also called the *adenohypophysis* or *pars distalis,* is sometimes called the "master gland" because of its role in regulating and maintaining other endocrine glands. Hormones that act on other endocrine glands are called *tropic hormones;* all the hormones produced in the anterior lobe are polypeptides. Two capillary beds connected by venules make up the *hypophyseal portal system,* which connects the hypothalamus with the anterior lobe.

Among the hormones produced in the anterior lobe of the pituitary gland are the following:

- **Follicle-stimulating hormone (FSH):** Signals an immature Graafian follicle in an ovary to mature into an ovum. After ovulation, the ovary forms the *corpus luteum,* producing the hormone estrogen. Negative feedback from the estrogen blocks further secretion of FSH. Guys, don't think you needn't worry about FSH: It's present in you, too, encouraging development and maturation of sperm. (For a review of the male and female reproductive systems, flip to Chapters 13 and 14.)

- **Luteinizing hormone (LH):** Stimulates formation of the yellow body, or corpus luteum, on the surface of the ovary after an ovum has been released. In men, LH stimulates the development of interstitial cells and fresh production of testosterone. It's also known as interstitial-cell stimulating hormone (ICSH).

- **Lactogenic hormone, or prolactin (PRL):** Promotes milk production in mammary glands, which are considered nonendocrine targets.

- **Thyrotropic hormone, or thyroid-stimulating hormone (TSH):** Controls the development and release of thyroid gland hormones *thyroxin* and *triiodothyronine.* The hypothalamus regulates TSH secretion by secreting thyrotropin-releasing hormone (TRH).

- **Adrenocorticotropic hormone (ACTH), or corticotropin:** Is a polypeptide composed of 39 amino acids that regulate the development, maintenance, and secretions of the cortex of the adrenal gland. Secreted by the anterior pituitary gland, it's an important component of the hypothalamus–pituitary–adrenal axis and often produced in response to stress, along with its precursor, corticotropin-releasing hormone from the hypothalamus.

- **Somatotropic hormone, or growth hormone (GH):** Stimulates body weight growth and regulates skeletal growth. This is the only hormone secreted by the anterior lobe that has a general effect on nearly every cell in the body (also regarded as nonendocrine targets).

The *posterior lobe,* or *neurohypophysis,* of the pituitary gland stores and releases neurohormone secretions produced by the hypothalamus. This lobe is connected to the hypothalamus by the *hypophyseal tract,* nerve axons with cell bodies lying in the hypothalamus. Whereas the adenohypophysis is made up of epithelial cells, the neurohypophysis is largely composed of modified nerve fibers and neuroglial cells called *pituicytes.*

Among the hormones produced in the posterior lobe of the pituitary gland are the following:

> ✔ **Oxytocin:** Stimulates contraction of the uterine smooth muscle during childbirth and the release of breast milk in nursing women. Experts believe that intimate social activities such as hugging, touching, selective social recognition, bonding, and sexual pleasures release oxytocin in both males and females.

> ✔ **Vasopressin, or antidiuretic hormone (ADH):** Constricts smooth muscle tissue in the blood vessels, elevating blood pressure and increasing the amount of water reabsorbed by the kidneys, which reduces the production of urine. The hypothalamus has special neurons called *osmoreceptors* that monitor the amount of solute in the blood.

See how much of this information you're absorbing:

0. The hormone that stimulates ovulation is the

    a. follicle-stimulating hormone (FSH).

    b. antidiuretic hormone (ADH).

    c. oxytocin.

    d. thyroid-stimulating hormone (TSH).

    e. luteinizing hormone (LH).

A. The correct answer is luteinizing hormone (LH). Don't be fooled into thinking it's FSH; that hormone does its job earlier, when it encourages a follicle to mature.

8.–12. Mark the statement with a T if it's true or an F if it's false:

    8. _____ The pituitary gland consists of two parts: an endocrine gland and modified nerve tissue.

    9. _____ The pituitary gland is found in the sella turcica of the temporal bone.

    10. _____ The adenohypophysis is called the master gland because of its influence on all the body's tissues.

    11. _____ ADH causes constriction of smooth muscle tissue in the blood vessels, which elevates the blood pressure.

    12. _____ The neurohypophysis stores and releases secretions produced by the hypothalamus.

13. Why is the hypothalamus considered to be just as important as the pituitary gland?

    a. It contracts and relaxes to regulate the so-called master gland.

    b. It brings in substances from the thyroid.

    c. It tells the pituitary gland what to release and when.

    d. It controls how large the thymus can grow.

    e. It influences the parathyroid more directly than the pituitary gland does.

14. Which of the following is not a pituitary hormone?

a. Prolactin

b. Follicle-stimulating hormone (FSH)

c. Growth hormone (GH)

d. Progesterone

e. Luteinizing hormone (LH)

The Supporting Cast of Glandular Characters

While the pituitary orchestrates the show at center stage, the endocrine system enjoys the support of a number of other important glands. Lying in various locations throughout the body, the glands in the following sections secrete check-and-balance hormones that keep the body in tune.

Topping off the kidneys: The adrenal glands

Also called *suprarenals,* the adrenal glands lie atop each kidney. The central area of each is called the *adrenal medulla,* and the outer layers are called the *adrenal cortex.* Each glandular area secretes different hormones. The cells of the cortex produce more than 30 steroids, including the hormones *aldosterone, cortisone,* and some sex hormones. The medullar cells secrete *epinephrine* (you may know it as *adrenaline*) and *norepinephrine* (also known as *noradrenaline*).

The cortex

Made up of closely packed epithelial cells, the adrenal cortex is loaded with blood vessels. Layers form an outer, middle, and inner zone of the cortex. Each zone is composed of a different cellular arrangement and secretes different steroid hormones.

- The *zona glomerulosa* (outer zone) produces aldosterone.
- The *zona fasciculata* (middle zone) secretes cortisone (also called cortisol).
- The *zona reticularis* (inner zone) secretes small amounts of *gonadocorticoids* or sex hormones.

The following are among the hormones produced by the cortex:

- *Aldosterone,* or *mineralocorticoid,* regulates electrolytes (sodium and potassium mineral salts) retained in the body. It promotes the conservation of water and reduces urine output.

- *Cortisone,* or *cortisol,* acts as an antagonist to insulin, causing more glucose to form and increasing blood sugar to maintain normal levels. Elevated levels of cortisone speed up protein breakdown and inhibit amino acid absorption.

- *Androgens* and *estrogen* are cortical sex hormones. Androgens generally convey anti-feminine effects, thus accelerating maleness, although in women adrenal androgens maintain the sexual drive. Too much androgen in females can cause *virilism* (male secondary sexual characteristics). Estrogen has the opposite effect, accelerating femaleness. Too much estrogen in a male produces feminine characteristics.

The medulla

The adrenal medulla is made of irregularly shaped *chromaffin cells* arranged in groups around blood vessels. The sympathetic division of the autonomic nervous system controls these cells as they secrete adrenaline and noradrenaline. The adrenal medulla produces approximately 80 percent adrenaline and 20 percent noradrenaline. Both hormones have similar molecular structure and physiological functions:

- **Adrenaline** accelerates the heartbeat, stimulates respiration, slows digestion, increases muscle efficiency, and helps muscles resist fatigue.

- **Noradrenaline** does similar things but also raises blood pressure by stimulating contraction of muscular arteries.

The terms "adrenaline" and "noradrenaline" are interchangeable with the terms "epinephrine" and "norepinephrine." You're likely to encounter both in textbooks and exams.

Thriving with the thyroid

The largest of the endocrine glands, the thyroid is shaped like a large butterfly with two lobes connected by a fleshy *isthmus* positioned in the front of the neck, just below the larynx and on either side of the trachea (you can see the thyroid in Figure 16-1). A transport mechanism called the *iodide pump* moves the iodides from the bloodstream for use in creating its two primary hormones, *thyroxin* and *triiodothyronine,* which regulate the body's metabolic rate. (More thyroxin is secreted than triiodothyronine.) *Extrafollicular cells* (also called *parafollicular* or *C cells*) secrete *calcitonin,* a polypeptide hormone that helps regulate the concentration of calcium and phosphate ions by inhibiting the rate at which they leave the bones. High blood calcium levels stimulate the secretion of more calcitonin.

Thyroxin (T_4) and triiodothyronine (T_3) regulate cellular metabolism throughout the body, but the thyroid needs iodine to manufacture those hormones. Iodine insufficiency causes the thyroid to swell in a condition called a *goiter.*

Pairing up with the parathyroid

The parathyroid consists of four pea-sized glands that are attached posteriorly to the thyroid gland secreting *parathormone,* also known as *parathyroid hormone* (PTH). This large polypeptide regulates the balance of calcium levels in the blood and bones as well as controls the rate at which calcium is excreted into urine. When blood calcium levels dip, the parathyroid secretes PTH, which increases calcium absorption from the intestine, decreases calcium excretion, increases phosphate excretion, removes calcium from the bones, and stimulates secretion of calcitonin by the thyroid C cells. Blood calcium ion homeostasis is critical to the conduction of nerve impulses, muscle contraction, and blood clotting.

Pinging the pineal gland

The pineal gland, also called the *epiphysis, epiphysis cerebri,* or *pineal body,* is a small, oval gland thought to play a role in regulating the body's biological clock. It lies between the cerebral hemispheres and is attached to the thalamus near the roof of the third ventricle.

Because it both secretes a hormone and receives visual nerve stimuli, the pineal gland is considered part of both the nervous system and the endocrine system. Its hormone *melatonin* is believed to play a role in circadian rhythms, the pattern of repeated behavior associated with the cycles of night and day, affecting the modulation of wake/sleep patterns and photoperiodic (seasonal) functions. The pineal gland is affected by changes in light, producing its highest levels of secretion at night and its lowest levels during daylight hours.

Thumping the thymus

As discussed in Chapter 11, the thymus is thought to secrete a group of peptides called *thymosin* that affect the production of lymphocytes (white blood cells). Thymosin promotes the production and maturation of T lymphocyte cells as part of the body's immune system. The gland is large in children and atrophies with age.

Pressing the pancreas

The pancreas is both an exocrine and an endocrine gland, which means that it secretes some substances through ducts while others go directly into the bloodstream. (We cover its exocrine functions in Chapter 9.) The pancreatic endocrine glands are clusters of cells called the *islets of Langerhans* (also known as *pancreatic islands* or *pancreatic islets*). Within the islets are a variety of cells, including

- **A cells (alpha cells)** that secrete the hormone *glucagon,* a polypeptide of 29 amino acids that increases blood sugar

- **B cells (beta cells)** that secrete *insulin,* a two-linked polypeptide chain of 21 amino acids that decreases blood sugar levels, increases lipid synthesis, and stimulates protein synthesis

- **D cells (delta cells)** that secrete *somatostatin,* a growth hormone–inhibiting factor that inhibits the secretion of insulin and glucagons

- **F cells (PP cells)** that secrete a pancreatic polypeptide that regulates the release of pancreatic digestive enzymes

Diabetes mellitus, or sugar diabetes, occurs when the islets of Langerhans cease producing insulin. Without insulin, the body can't metabolize sugar, so it builds up in the blood and is secreted by the kidneys.

See whether all this information has your hormones raging:

15.–19. Mark the statement with a T if it's true or an F if it's false:

15. _____ The adrenal glands are located in the cortex of the kidneys.

16. _____ Adrenaline is functional in the absorption of stored carbohydrates and fat.

17. _____ Aldosterone is functional in regulating the amount of insulin in the body.

18. _____ The sympathetic division of the autonomic nervous system controls the cells of the adrenal medulla.

19. _____ The layers of the adrenal medulla form outer, middle, and inner zones.

20. The thymus produces thymosin, which

    a. stimulates the pineal body.

    b. reduces the pituitary gland.

    c. initiates lymphocyte development.

    d. controls the hypothalamus.

21. Which of the following does the hormone aldosterone affect?

    a. Regulation of body fluid concentration

    b. Removal of wastes

    c. The fight-or-flight response

    d. Emotional outbursts

22.–26. Mark the statement with a T if it's true or an F if it's false.

    22. _____ Iodine is a necessary component of thyroxin (T_4) and triiodothyronine (T_3).

    23. _____ Follicular cells of the thyroid produce hormones that affect the metabolic rate of the body.

    24. _____ A transport mechanism called the sodium pump moves the iodides into the follicle cells.

    25. _____ Thyroxin (T_4) is normally secreted in lower quantity than triiodothyronine (T_3).

    26. _____ The hormone calcitonin helps regulate the concentration of sodium and potassium.

27. Which statement is not true of the pineal gland?

    a. It secretes melatonin.

    b. Nerve fibers stimulate the pineal cells.

    c. As light decreases, secretion increases.

    d. It promotes immunity.

28.–32. Mark the statement with a T if it's true or an F if it's false:

    28. _____ The parathyroid gland contains cells that secrete parathormone or parathyroid hormone (PTH).

    29. _____ Melatonin is a polypeptide that regulates the balance of calcium in the blood and bones.

    30. _____ The pineal gland responds to light, producing higher levels of secretions at night than during the day.

    31. _____ Thymosin promotes the production and maturation of erythrocyte cells.

    32. _____ The parathyroid hormone can prompt calcium to move from bone.

33. The adrenal medulla serves one primary function. What is it?

    a. To produce erythrocyte cells

    b. To absorb excess iodine from the bloodstream

    c. To produce epinephrine

    d. To secrete polypeptides

34. Clusters of cells can be found within the islets of Langerhans in the pancreas. Which variety of cell secretes glucagon to increase blood sugar?

    a. Beta cells, also known as B cells

    b. Alpha cells, also known as A cells

    c. PP cells, also known as F cells

    d. Delta cells, also known as D cells

35.–45. Use the terms that follow to identify the structures of the endocrine system shown in Figure 16-2.

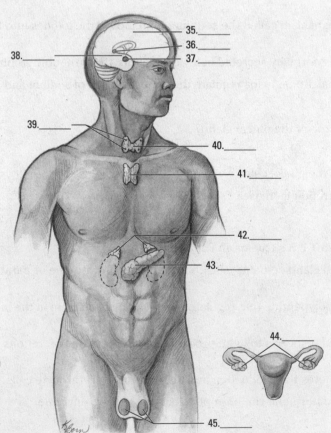

35._____
36._____
37._____
38._____
39._____
40._____
41._____
42._____
43._____
44._____
45._____

Figure 16-2:
The endo-
crine
system.

Illustration by Kathryn Born, MA

a. Thyroid gland

b. Pineal gland

c. Pituitary gland

d. Adrenal or suprarenal glands

e. Ovaries

f. Parathyroid gland

g. Testes

h. Hypothalamus

i. Pancreas

j. Brain

k. Thymus

Dealing with Stress: Homeostasis

Nothing upsets your delicate cells more than a change in their internal environment. A stimulus such as fear or pain provokes a response that upsets your body's carefully maintained equilibrium. Such a change initiates a nerve impulse to the hypothalamus that activates the sympathetic division of the autonomic nervous system and increases secretions from the adrenal glands. This change — called a *stressor* — produces a condition many know all too well: *stress*. Stress is anything that disrupts the normal function of the body. The body's immediate response is to push for *homeostasis* — keeping everything the same inside. Inability to maintain homeostasis — a condition known as *homeostatic imbalance* — may lead to disease or even death.

The body's effort to maintain homeostasis invokes a series of reactions called the *general stress syndrome* that's controlled by the hypothalamus. When the hypothalamus receives stress information, it responds by preparing the body for fight or flight — in other words, some kind of decisive, immediate, physical action. This reaction increases blood levels of glucose, glycerol, and fatty acids; increases the heart rate and breathing rate; redirects blood from skin and internal organs to the skeletal muscles; and increases the secretion of adrenaline from the adrenal medulla. The hypothalamus releases *corticotropin-releasing hormone* (CRH) that stimulates the anterior lobe of the pituitary to secrete *adrenocorticotropic hormone* (ACTH), which tells the adrenal cortex to secrete more cortisone. That cortisone supplies the body with amino acids and an extra energy source needed to repair any injured tissues that may result from the impending crisis.

As part of the general stress syndrome, the pancreas produces glucagon, and the anterior pituitary secretes growth hormones, both of which prepare energy sources and stimulate the absorption of amino acids to repair damaged tissue. The posterior pituitary secretes antidiuretic hormone, making the body hang on to sodium ions and spare water. The subsequent decrease in urine output is important to increase blood volume, especially if there is bleeding or excessive sweating.

Wow. With the body gearing up like that every time, it's no wonder that people subjected to repeated stress are often sickly.

We'll try not to stress you out with these practice questions:

46.–50. Mark the statement with a T if it's true or an F if it's false.

46. _____ The hypothalamus controls reactions to combat general stress syndrome.

47. _____ General stress syndrome is controlled by the thymus.

48. _____ During stress, the pancreas produces thyroxin (T_4).

49. _____ Homeostatic imbalance has primarily positive effects on the body.

50. _____ Cortisone released in reaction to stress provides an extra energy source for tissue repair.

51. When changes occur in the body's internal environment, a hormonal chain reaction occurs, beginning with the _____, which prompts secretions from the _____, which in turn instruct the _____ to secrete cortisone.

   a. hypothalamus; thymus; pituitary

   b. pituitary; hypothalamus; thymus

   c. hypothalamus; pituitary; adrenal cortex

   d. adrenal medulla; hypothalamus; pituitary

52. Stress activates a set of body responses called the general stress syndrome that includes all of the following except

   a. increases in cardiac output.

   b. changes in blood pressure.

   c. increased sweat.

   d. release of additional insulin.

53. The body's initial reaction to a stressor is

   a. the fight-or-flight response.

   b. the repair response.

   c. to promote rapid wound healing.

   d. the stress reflex.

   e. to promote normal metabolism.

Answers to Questions on the Endocrine System

The following are answers to the practice questions presented in this chapter.

1. The endocrine system brings about changes in the metabolic activities of the body tissue. **True.** Metabolism is one of the areas influenced by hormones.

2. The amount of hormone released is determined by the body's need for that hormone at the time. **True.** In many ways, it's a self-regulating system: Just enough hormones are distributed to balance everything else out.

3. The glands of the endocrine system are composed of cartilage cells. **False.** That connection makes no sense whatsoever.

4. Endocrine glands aren't functional in reproductive processes. **False.** The endocrine system is a key component in reproduction.

5. Some hormones can be derivatives of amino acids, whereas others are synthesized from cholesterol. **True.** Amino acids work for nonsteroid hormones, and cholesterol does the trick for steroid-based hormones.

6. **a. Exocrine** glands secrete their product through ducts while **endocrine** glands secrete their product into the interstitial fluid, which flows into the blood.

7. Some hormones affect a cell by binding to membrane receptors, while others diffuse into the cell. Hormones that bind to membrane receptors are called **b. water-soluble,** and hormones that diffuse into the cell are called **lipid-soluble.**

8. The pituitary gland consists of two parts: an endocrine gland and modified nerve tissue. **True.** Remember that the anterior lobe is mostly epithelial cells, whereas the posterior lobe contains primarily nerve cells.

9. The pituitary gland is found in the sella turcica of the temporal bone. **False.** The pituitary gland is in the sphenoid bone, not the temporal bone.

10. The adenohypophysis is called the master gland because of its influence on all the body's tissues. **False.** It earned the title "master gland" because of its influence over the other endocrine glands.

11. ADH causes constriction of smooth muscle tissue in the blood vessels, which elevates the blood pressure. **True.** It's understandable that constriction increases pressure.

12. The neurohypophysis stores and releases secretions produced by the hypothalamus. **True.** Seems a strange thing for a structure made of nerve cells to do, but it does its job well.

13. Why is the hypothalamus considered to be just as important as the pituitary gland? **c. It tells the pituitary gland what to release and when.** If you remember that the pituitary gland is attached to the bottom of the hypothalamus, you can see the importance of the relationship between the two.

14. Which of the following is not a pituitary hormone? **d. Progesterone.** That's made by the corpus luteum in the ovary following ovulation.

15. The adrenal glands are located in the cortex of the kidneys. **False.** They're atop the kidneys.

16. Adrenaline is functional in the absorption of stored carbohydrates and fat. **False.** Adrenaline does lots of things, but not that.

17   Aldosterone is functional in regulating the amount of insulin in the body. **False.** This hormone regulates mineral salts.

18   The sympathetic division of the autonomic nervous system controls the cells of the adrenal medulla. **True.** Those cells are busy secreting both adrenaline and noradrenaline.

19   The layers of the adrenal medulla form outer, middle, and inner zones. **False.** The layers of the adrenal cortex form those three zones.

20   The thymus produces thymosin, which **c. initiates lymphocyte development.**

21   Which of the following does the hormone aldosterone affect? **a. Regulation of body fluid concentration.** Aldosterone is just one of more than 30 steroids produced by the adrenal cortex.

22   Iodine is a necessary component of thyroxin (T_4) and triiodothyronine (T_3). **True.** The body can't make those hormones without iodine.

23   Follicular cells of the thyroid produce hormones that affect the metabolic rate of the body. **True.** They do it by concentrating iodide.

24   A transport mechanism called the sodium pump moves the iodides into the follicle cells. **False.** Don't let the thyroid's iodide pump make you think it also has a sodium pump. It doesn't.

25   Thyroxin (T_4) is normally secreted in a lower quantity than triiodothyronine (T_3). **False.** In fact, it's just the opposite — more T_4 is secreted than T_3.

26   The hormone calcitonin helps regulate the concentration of sodium and potassium. **False.** Actually, calcitonin lowers plasma calcium and phosphate levels.

27   Which statement is not true of the pineal gland? **d. It promotes immunity.** It's more of a biological-clock kind of gland.

28   The parathyroid gland contains cells that secrete parathormone or parathyroid hormone (PTH). **True.** This regulates the balance of calcium levels in the blood and bones.

29   Melatonin is a polypeptide that regulates the balance of calcium in the blood and bones. **False.** Melatonin is thought to regulate circadian rhythms.

30   The pineal gland responds to light, producing higher levels of secretions at night than during the day. **True.** That's why it's also considered part of the nervous system.

31   Thymosin promotes the production and maturation of erythrocyte cells. **False.** Thymosin works on lymphocytes.

32   The parathyroid hormone can prompt calcium to move from bone. **True.** After all, the bones are mineral reservoirs.

33   The adrenal medulla serves one primary function. What is it? **c. To produce epinephrine.** That's about all the adrenal medulla does. (Actually, it produces 80 percent epinephrine and 20 percent norepinephrine.)

34   Clusters of cells can be found within the islets of Langerhans in the pancreas. Which variety of cell secretes glucagon to increase blood sugar? **b. Alpha cells, also known as A cells.** The secretion from A cells increases blood sugar, insulin from B cells decreases blood sugar, somatostatin from D cells inhibits both insulin and glucagon, and the polypeptide from F cells regulates digestive enzymes.

35–45   Following is how Figure 16-2, the endocrine system, should be labeled.

35. **j. Brain;** 36. **b. Pineal gland;** 37. **c. Pituitary gland;** 38. **h. Hypothalamus;** 39. **f. Parathyroid gland;** 40. **a. Thyroid gland;** 41. **k. Thymus;** 42. **d. Adrenal or suprarenal glands;** 43. **i. Pancreas;** 44. **e. Ovaries;** 45. **g. Testes.**

46 The hypothalamus controls reactions to combat general stress syndrome. **True.** It does so by preparing the body for fight or flight.

47 General stress syndrome is controlled by the thymus. **False.** That would be the hypothalamus, actually.

48 During stress, the pancreas produces thyroxin (T_4). **False.** That's the thyroid's job, and T_4 has nothing to do with stress, anyway.

49 Homeostatic imbalance has primarily positive effects on the body. **False.** Imbalance is a bad thing — so bad that it's capable of killing.

50 Cortisone released in reaction to stress provides an extra energy source for tissue repair. **True.** It's part of how the body gears up to manage an impending crisis.

51 When changes occur in the body's internal environment, a hormonal chain reaction occurs, beginning with the **c. hypothalamus,** which prompts secretions from the **pituitary,** which instruct the **adrenal cortex** to secrete cortisone.

52 Stress activates a set of body responses called the general stress syndrome that includes all of the following except **d. release of additional insulin.** One of the body responses activated during stress occurs in the pancreas, where glucagon production goes up. Insulin would counteract that.

53 The body's initial reaction to a stressor is **a. the fight-or-flight response.** You're either going to put up your dukes or run like the wind.

Part VI

The Part of Tens

In this part . . .

- ✔ Brush up on the ten best ways to study anatomy and physiology effectively.

- ✔ Explore ten fun physiology facts in a free-wheeling testament to trivia.

Chapter 17

Ten Study Tips for Anatomy and Physiology Students

....

In This Chapter

▶ Playing with words and outlines

▶ Getting the most out of study groups

▶ Tackling tests and moving forward

....

*W*hat's the best way to tackle anatomy and physiology and come out successful on the other side? Of course, a good memory helps plenty — after all, you have to remember what goes where and which terminology attaches to which part. But with a little advance planning and tricks of the study trade, even students who complain that they can't remember their own names on exam day can summon the right terminology and information from their scrambled synaptic pathways. In this chapter, we cover ten key things you can start doing today to ensure success not only in anatomy and physiology but in any number of other classes.

Writing Down Important Stuff in Your Own Words

This is a simple idea that far too few students practice regularly. Don't stop at underlining and highlighting important material in your textbooks and study guides: Write it down. Or type it up. Whatever you do, don't just regurgitate it exactly as presented in the material you're studying. Find your own words. Create your own analogies. Tell your own tale of what happens to the bolus as it ventures into the digestive tract. Detail the course followed by a molecule of oxygen as it enters through the nose. Draw pictures of the differences between meiosis and mitosis. When you're answering practice questions, pay special attention to the ones you get wrong. Write reflections about why you answered incorrectly and what you need to remember about the right answer. Completely relax into the process with the knowledge that no one else ever has to see what you write, type, or sketch. All anyone else will ever see is your successful completion of the course!

Gaining Better Knowledge through Mnemonics

Studying anatomy and physiology involves remembering lists of terms, functions, and processes. Sprinkled throughout this book are suggestions to take just the first letter or two of each word from a list to create an acronym. Occasionally, we help you go one step beyond the acronym to a clever little thing called a *mnemonic device*. Simply put, the mnemonic is the thing you commit to memory as a means for remembering the more technical thing for which it stands. For example, a question in Chapter 7 asks you to list in order the epidermal layers from the dermis outward; we suggest that you commit the following phrase to memory: Be Super Greedy, Less Caring. Just like that, a complicated list like basale, spinosum, granulosum, lucidum, corneum gets a little closer to a permanent home in your brain.

Not feeling terribly clever at the moment you need a useful mnemonic? Surf on over to www.medicalmnemonics.com, which touts itself as the world's database of these useful tools. Here's a sampling of the site's offerings:

✔ To remember the three types of tonsils: "PPL (people) have tonsils: **P**haryngeal, **P**alatine, and **L**ingual."

✔ To remember the spleen's location and dimensions: "Count 1, 3, 5, 7, 9, 11: The spleen is 1 inch by 3 inches by 5 inches, weighs 7 ounces, and underlies ribs 9 through 11."

✔ To remember the cranial bones: "Think PEST OF 6: **P**arietal, **E**thmoid, **S**phenoid, **T**emporal, **O**ccipital, **F**rontal."

✔ To remember the path out of the body from the top of the intestines: "Dow Jones Industrial Average Closing Stock Report: **D**uodenum, **J**ejunum, **I**leum, **A**ppendix, **C**olon, **S**igmoid, **R**ectum."

Discovering Your Learning Style

Every person has his or her own sense of style, and woe betide anyone who tries to shoehorn the masses into a single style. The same, of course, is true of students. To get the most out of your study time, you need to figure out what your learning style is and alter your study habits to accommodate it. No idea what we're talking about? Answer the questionnaire posted at www.vark-learn.com/english/page.asp?p=questionnaire, and the VARK guide to learning styles will tell you more about yourself than your last psychotherapist.

VARK, as you may have suspected, is an acronym. It stands for Visual (learning by seeing), Aural (learning by hearing), Reading/Writing (learning by reading and writing), and Kinesthetic (learning by touching, holding, or feeling). If you're a visual learner, you may get more out of anatomy and physiology by seeing the real thing in the flesh. If you're an aural learner, you may learn best in the classroom as the teacher lectures. If you're a reading and writing kind of learner, you'll get the most out of our first tip to write stuff down. And if you're a kinesthetic learner, there's nothing like touching or holding to commit something to memory.

Getting a Grip on Greek and Latin

If you keep thinking "It's all Greek to me," congratulations on your insight! The truth of the matter is that most of it actually *is* Greek. So dust off your foreign language learning skills and begin with the basic vocabulary of medical terminology. Get started with the Greek and Latin roots, prefixes, and suffixes that appear on this book's Cheat Sheet at www.dummies. com/cheatsheet/anatomyphysiologywb. You'll soon discover that for every "little" word you learn, a whole mountain of additional terms and phrases are just waiting to be discovered.

Connecting with Concepts

It happens time and again in anatomy and physiology: One concept or connection mirrors another yet to be learned. But because you're focusing so hard on this week's lesson, you lose sight of the value in the previous month's lessons. For example, a concept like metabolism comes up in a variety of ways throughout your study of anatomy and physiology. When you encounter a repeat concept like that, create a special page or two for it at the back of your notebook or link the concept to a separate computer file. Then, every time the term comes up in class or in your textbook, add to the running list of notes on that concept. You'll have references to metabolism at each point it comes up *and* you'll be able to analyze its influences across different body systems.

Forming a Study Group

If you're really lucky, someone in your class (or maybe it's even you) has already suggested forming that time-honored tradition — a *study group*. The power of group members to fill gaps in your knowledge is priceless. But don't restrict it to late-night cramming just before each test. Meet with your group at least once a week to go over lecture notes and textbook readings. If it's true that people only retain about 10 percent of what they hear or read, then it makes sense that your fellow group members will recall things that slipped immediately from your mind.

Outlining What's to Come

As you read through a chapter of your textbook to prepare for the next lecture, prepare an outline of what you're reading, leaving plenty of space between subheadings. Then, during the lecture, take your notes within the outline you've already created. Piecing together an incomplete puzzle shows you where the key gaps in your knowledge may be.

Putting In Time to Practice

Flash cards, mnemonic drills, practice tests — be creative and practice, practice, practice! The more you know about the format of any upcoming exam, the better. Sometimes instructors share tidbits about what they plan to emphasize, but sometimes they don't. In the end, if you've done the work and put in the time to study and practice with information outside of class, the exact structure and content of an exam shouldn't make much difference.

Sleuthing Out Clues

Okay, it's test time! Take advantage of the test itself. You may find that the answer to an exam question that stumps you is revealed — at least partially — in the phrasing of a subsequent question. Stay alert to these blessed little gifts even when you think that you already understand all the anatomical structures and physiological processes. You won't be the first student to change an answer after working your way through an exam.

Reviewing Your Mistakes

The test is done and the grades are in. So there was a really tough question or two on the test and you blew it big-time? It's hardly a missed opportunity — this is where rolling with the punches really pays off. Go back over the entire test and pay extra attention to what you got wrong. Start your next practice sessions with those questions, and stay alert for upcoming material that may trip you up in a similar way.

Chapter 18

Ten Fun Physiology Facts

Basic anatomy may be fairly straightforward, but how the human body uses all those parts can present a smorgasbord of interesting discoveries. In this chapter, we give you just a peek at ten of the more intriguing aspects of what makes the body tick.

Boning Up on the Skeleton

Make no bones about it — the human skeleton is a trove of trivia. Try a few of these on for size (and check out Chapter 5 for the scoop on the skeleton):

✔ People are born with 300 bones in their infant bodies, but by the time they're adults they only have 206. Why? Because human bodies spend infancy and childhood putting the finishing touches on their skeletons, knitting together two or more bones into one. The only bone fully grown at birth is in the ear. Plus, over a period of about seven years, each bone in the body is slowly replaced until it is a new bone.

✔ You're taller in the morning. No, really, you are. After you stand up from bed and get your day going, the cartilage between your back bones becomes compressed over time. By the end of the day, you're about one centimeter shorter.

✔ Bones make up about 14 percent of an average individual's weight, yet they're as strong as granite. A block of bone the size of a matchbox can support 9 tons — four times more weight than concrete.

✔ Bones can self-destruct. Excessive exposure to cadmium can prompt them to do so by triggering premature *apoptosis,* the controlled death of cells that takes place as part of normal growth and development. Alternatively, if you're not taking in enough calcium, certain hormones can leach calcium from your own bones to try to balance the blood's supply of this essential mineral.

Flexing Your Muscles

Besides your heart, the strongest muscle in the body (compared to its size) is your tongue. Hey, something has to counterbalance those 200 pounds of force the jaw muscles can deliver when you're chewing. And which muscles move more each day than even your heart? The ones surrounding your eyes, called the extraocular muscles. At three moves per second, they clock in more than 100,000 motions per day.

Every time you take a step, you employ around 200 of the roughly 600 muscles in your body. Researchers estimate that every minute you spend walking extends your life by anywhere from 90 seconds to two minutes. No wonder doctors advise that people shoot for 10,000 steps each day!

Thinking you'll just lift weights instead? Choose free weights over using a weight machine. The action of balancing the weights yourself rather than letting a machine do it for you builds muscle mass faster. (Chapter 6 introduces you to the muscles.)

Fighting Biological Invaders

As the body's internal defense network, the immune system sets blood and lymph against biological invaders seeking to colonize you. But its first line of defense is also the body's largest organ: the skin. If your skin — and some of the "friendly" bacteria living there — didn't secrete natural antibacterial compounds, you could wake up in the morning coated in microbial slime. As it is, though, most of the pathogens that land on you die quickly. (See Chapter 7 for details on the skin.)

The skin isn't the only thing working against the invading microbial hordes; the enzyme lysozyme found in human tears, saliva, and mucus is custom-made to disintegrate bacterial cell walls. (Remember: As eukaryotes, people's cells don't have walls).

Not that all microorganisms are bad. You have between 2 and 5 pounds of bacteria living inside you, much of it in the intestines. As scientists have begun to understand what that microbial life is up to, it has become clear that your internal "microbiome" is a big part of what keeps you healthy.

Dissolving into Dust

Every human alive today spent about 30 minutes at the start as a single cell. Now, however, your body is making 25 million new cells every day, and you shed and regrow all of your outer skin cells about every 27 days. (See Chapter 7 for more about the skin.) Beards are the fastest-growing hairs on the human body; if an average man never shaved or trimmed the growth, he would die with nearly 30 feet of beard hanging off his chin. The average human scalp has 100,000 hairs — blondes have more hair, on average, than dark-haired people do — and you lose between 40 and 100 strands of hair each day. Fingernails grow nearly four times faster than toenails do.

That's quite a bit of cellular production going on. Ever wonder what happens to those cells when they die? After all, you shed around one and a half pounds of dead skin every year.

It's dust — common household dust. Most of the dust you brush around in your house is dead skin. If you think that's gritty, consider this: The ashes of a cremated body weigh around 9 pounds.

Swallowing Some Facts about Saliva and the Stomach

Besides being a digestive kick-starter, saliva prevents tooth decay and keeps your throat and mouth from drying out. It may not feel like it, but those six little glands make nearly half a gallon — about 1.5 liters — of saliva every day. Over the course of a lifetime, that's enough to fill about two average-sized swimming pools.

Hard to stomach? Consider this: With hydrochloric acid inside that's so corrosive it can dissolve wood and steel, your stomach should consume itself. But it doesn't. Why? Because you make your own natural supply of antacid. The epithelial cells lining your stomach secrete a steady supply of bicarbonate that neutralizes stomach acid on contact.

Be forewarned, though: That soothing bicarbonate produces carbon dioxide, among other things, as a by-product. That has to go somewhere. Some of it comes up in the form of a belch. And some of it contributes to the 17 ounces per day of flatulence that the average healthy human releases. (See Chapter 9 for an introduction to the digestive system.)

Appreciating the Extent of the Circulatory System

We tell you in Chapter 10 about the heart, the hardest-working muscle in the entire body, and its web of liquid connective tissue more commonly known as blood. But we don't mention a few facts that may make you appreciate your circulatory system even more.

In one day, the average individual's heart exerts enough power to lift a 1-ton weight more than 40 feet off the ground. During that day's pumping, the body's red blood cells travel about 12,000 miles, or roughly half the distance around the Earth at the equator. They've got plenty of room to roam; if you laid every blood vessel in a body end-to-end, they would stretch roughly one-quarter of the way to the moon, or about 60,000 miles.

If that's hard to believe, consider this: Every square inch of skin contains 20 feet of blood vessels. Tissue the size of a pinhead contains 2,000 to 3,000 capillaries. You certainly have plenty of blood to fill all that space; you have 2.5 trillion red blood cells, more or less, in your body at any given moment, and your bone marrow creates 100 billion new ones every single day. That's so much blood that it would take 1.2 million mosquitoes each drinking their fill once to completely drain the average human of blood.

Finding a Surprising Link between Allergies and Cancer

Do you feel a sneeze coming on the moment you even hear the word "pollen"? Allergies may feel like they simply cannot have an upside, but don't wish your hay fever away too fast: Two decades of studies suggest your suffering may not be entirely for naught.

When allergies push the immune system into overdrive, they may be doing you a favor — even as they make you miserable. Allergy sufferers are less likely to be diagnosed with cancer, particularly in tissues exposed to the outside environment (skin, lungs, mouth, throat and the gastrointestinal tract, among others). There are two schools of thought on why:

- ✔ Allergies prompt the production of those T cells we mention in Chapter 11. And T cells can destroy cancer cells. That's one theory.

- ✔ Another theory is that allergic reactions serve to engulf potential pathogens in mucous and expel them from the body quickly.

Looking at a Few of Your Extra Parts

With so many industrious components keeping you moving through your life, it can be startling to think about the number of body parts that, frankly, you just don't need. The first ones that come to mind often are the appendix and the wisdom teeth. The appendix, which does produce a few white blood cells, generally gets to stay put so long as it doesn't get infected. But that third set of molars does little but cause pain and push around the teeth you really need.

You also have a "tailbone," or coccyx, which serves no function other than as a high-scoring Scrabble word, and it's a painful reminder of hard falls on your backside. It's a collection of fused vertebrae that researchers believe are what's left of the mammalian tail our evolutionary ancestors once sported. (See Chapter 5 for more about bones.)

People have a few other vestigial organs. Men, believe it or not, have an undeveloped vestigial uterus hanging on one side of the prostate gland, while women have a vestigial vas deferens — a cluster of tubules near the ovaries that would have become sperm ducts if they had inherited a Y chromosome from their fathers. (Part IV has more details on both the male and female reproductive systems.)

Understanding Your Brain on Sleep

Superlatives about the human body tend to center around the brain. Fastest, neediest, most powerful — it is, after all, what makes us human. That 3-pound hunk of tissue demands 20 percent of the oxygen and calories the body takes in, communicates with 45 miles of nerves in the skin, contains individual neurons that can live more than a century, and generates enough energy when a person is awake — between 10 and 23 watts — to power a light bulb.

But the fun really begins when humans go to sleep. Scientists long have wondered why the body requires that people spend one-third of their lives unconscious. Although an individual may look quiet when she's with the sand man, studies have shown that the brain is more active in sleep than awake. Sleep, and the process of dreaming, helps the brain consolidate memories in ways that researchers are only beginning to understand.

Studies released in 2013 indicate that the brain is up to something else when humans sleep: It's taking out the trash. The lymphatic system that moves cellular waste out of the body doesn't reach into the brain. Instead, the brain is bathed in cerebrospinal fluid. When you sleep, your *glial cells* actually shrink up to 60 percent to widen the channels and sweep toxins away more efficiently. (Derived from the Greek word for "glue," glial cells are the connective, supporting cells in the nervous system.) Now scientists are studying this newly discovered *glymphatic system* to see whether it can be brought to bear in the battle against Alzheimer's and other forms of dementia. (Flip to Chapter 15 for full details on the brain.)

Getting Sensational News

When people talk about sensation, most of the time they're referring to the five primary senses — vision, hearing, taste, touch, and smell (see Chapter 15). But the notion of a "sixth sense" may be more than the stuff of science fiction. Vision combines senses for both light and color, and there is growing evidence that your *proprioception* — the ability to detect your relative position in space — may rely on your body's ability to detect magnetic fields in much the same way migrating birds do. Blind people have been known to develop echolocation, or the ability to hear subtle changes in sound bouncing back from otherwise unseen objects. And stress sometimes causes people to experience time dilation.

In fact, your senses are more subjective than you may like to admit. Things get even more complex when you consider the condition known as *synesthesia,* in which a person can "hear" color or "see" sound. The eyes, which can distinguish up to one million colors, routinely take in more information than the largest telescope ever created. The nose can identify and the brain subsequently can remember more than 50,000 smells. Static touch can discern an object about twice the diameter of an eyelash, while dynamic touch — dragging a finger along a surface — can detect bumps the size of a very large molecule.

Index

• *M* •

● **T** ●

• U •

• V •

• W •

• Z •

About the Authors

Janet Rae-Dupree has been covering innovation, science, and technology in Silicon Valley since 1993 for a number of publications, including the *New York Times, U.S. News & World Report, BusinessWeek,* and the *San Jose Mercury News.* She was a frequent guest on cable channel Tech TV's "Silicon Spin" technology talk show, and she was part of the Pulitzer Prize–winning team at the *Los Angeles Times* covering the city's riots in 1992. During the 2005–2006 academic year, Janet was a John S. Knight Journalism Fellow at Stanford University, where she studied human biology and researched innovation and market transfer. She freelances for various publications, working from her home in Half Moon Bay, California, where she lives with husband, Dave Dupree, and their son, Matthew (although two cats actually rule the roost).

Pat DuPree taught anatomy/physiology, biology, medical terminology, and environmental science for 24 years at several colleges and universities in Los Angeles County. She holds two undergraduate life science degrees and a master's degree from Auburn University and conducted cancer research at Southern Research Institute in Birmingham, Alabama, before joining the Muscogee Health Department in Columbus, Georgia. In 1970, she moved to Redondo Beach, California, where she was a university instructor and raised her two sons, Dave Dupree and Mark DuPree. Now Pat is retired and lives on lovely Pine Lake in rural Georgia with her husband of 55 years, Dr. James E. DuPree.

Dedication

To our loving family — Dave Dupree, Matthew Dupree, and Jim DuPree — for their patience and understanding as we pulled this book together and put so many other aspects of our lives on hold. To Dave, for being Mr. Fix-It on balky hard drives, recalcitrant computers, and fussy printers while supervising a fledgling Eagle Scout toward his goals; and to Matthew, for seizing the opportunity to make his way independently through coursework at Stanford University Online High School while skillfully navigating the rocky rapids of adolescence.

And especially from Pat to Jim for his love, enthusiastic support, assistance, encouragement, and computer skills, without which she could not have finished this workbook.

Authors' Acknowledgments

We owe gratitude to so many people, but this project has been first and foremost a DuPree family affair. We will never be able to thank our spouses enough for their patience and understanding from beginning to end. Dr. James E. DuPree was a constant helpmate to Pat on the East Coast, providing research, editing, computer assistance, and graphics support as well as priceless cheerleading and enthusiasm. Their son David E. Dupree filled the cheerleader role on the West Coast, keeping the computers and home network up and running, Janet focused and on track, and the gaping child-rearing gaps filled while she pushed to get the book done in what surely must be record time. And we can't forget to thank son/grandson Matthew J. Dupree for his stalwart patience while Mom and Gran-mom focused their attentions on the book instead of him.

We would also like to thank our agent, Matt Wagner, for his tireless efforts, as well as the many devoted people at John Wiley & Sons, Inc.

Publisher's Acknowledgments

Executive Editor: Lindsay Sandman Lefevere

Senior Project Editor: Georgette Beatty

Copy Editor: Christine Pingleton

Project Manager: Michelle Hacker

Technical Editors: Steve Dougherty, Scott Houser

Art Coordinator: Alicia B. South

Project Coordinator: Emily Benford

Illustrator: Kathryn Born, MA

Cover Image: iStock.com/Henrik5000